ARCTIC SILK ROADS

Studies in the Circumpolar North

Series Editors: Olga Ulturgasheva and Elizabeth Marino

The Circumpolar North encapsulates all the major issues confronting the world today: enduring colonial legacies for indigenous people and the landscape, climate change and resource extraction industries, international diplomatic tensions, and lived realities of small communities in the interconnected modern world system. This book series provides a showcase for cutting-edge academic research on the lives of Arctic and Sub-arctic communities past and present. Understanding the contemporary Circumpolar North requires a multiplicity of perspectives and we welcome works from the social sciences, humanities, and the arts.

Volume 7
Arctic Silk Roads: An Anthropology of the Unbuilt
Edited by Natalia Magnani and Matthew Magnani

Volume 6
Risky Futures: Climate, Geopolitics and Local Realities in the Uncertain Circumpolar North
Edited by Olga Ulturgasheva and Barbara Bodenhorn

Volume 5
Arctic Abstractive Industry: Assembling the Valuable and Vulnerable North
Edited by Arthur Mason

Volume 4
An Urban Future for Sápmi?: Indigenous Urbanization in the Nordic States and Russia
Edited by Mikkel Berg-Nordlie, Astri Dankertsen and Marte Winsvold

Volume 3
Urban Sustainability in the Arctic: Measuring Progress in Circumpolar Cities
Edited by Robert W. Orttung

Volume 2
Sustaining Russia's Arctic Cities: Resource Politics, Migration, and Climate Change
Edited by Robert W. Orttung

Volume 1
Leaving Footprints in the Taiga: Luck, Spirits and Ambivalence among the Siberian Orochen Reindeer Herders and Hunters
Donatas Brandišauskas

Arctic Silk Roads

An Anthropology of the Unbuilt

Edited by
Natalia Magnani and Matthew Magnani

berghahn
NEW YORK • OXFORD
www.berghahnbooks.com

First published in 2026 by

Berghahn Books
www.berghahnbooks.com

Library of Congress Cataloging-in-Publication Data
A C.I.P. cataloging record is available from the Library of Congress
Library of Congress Cataloging in Publication Control Number: 2025038560

British Library Cataloguing in Publication Data
A catalogue record for this book is available from the British Library

EU GPSR Authorized Representative
LOGOS EUROPE, 9 rue Nicolas Poussin, 17000, LA ROCHELLE, France
Email: Contact@logoseurope.eu

ISBN 978-1-83695-308-1 hardback
ISBN 978-1-83695-309-8 epub
ISBN 9781-83695-310-4 web pdf

https://doi.org/10.3167/9781836953081

The electronic open access publication of *Arctic Silk Roads: An Anthropology of the Unbuilt* has been funded by the Norwegian Research Council under a CC BY-NC-ND 4.0 license.

Contents

Illustrations

Figures

Map

Acknowledgments

We thank the many people whose voices, and innovative strivings for a better future, helped shape the chapters of this volume. The book project, including research and collaboration for many of the chapters, was funded by the Norwegian Research Council.

Introduction

Imagining Unbuilt Infrastructures across the Circumpolar North

Natalia Magnani and Matthew Magnani

The circumpolar North has long captured the global imagination as a staging ground for state interests. For much longer, the region has been a place of human dwelling, exchange, and movement. As their paths crossed and collided over the last centuries, the lives of northern inhabitants have entangled with colonial desires to explore, map, and exploit, from the search for the Northeast and Northwest Passages to connect trade, to scientific expeditions that documented the geology, flora, fauna, and not least the cultures of these regions in the process of acquisition and governance by southern centers of power.

At the nexus of climate change and geopolitics, conflicting powers are continuing to carve the region physically and in the anticipatory imagination. Melting sea ice is being reenvisioned as shipping routes and connected land corridors, which promise to streamline trade, transport, and resource extraction. Harnessing these visions, the region has been recast as a "Polar Silk Road" in China's Arctic policy and Belt and Road Initiative, and as a "Northern Sea Route" in Russian waters. The imaginaries exist in tension with NATO nuclear submarines and military exercises amid growing American and European security concerns in the Arctic following Russia's invasion of Ukraine. As Frank Billé (2020) and the authors of *Voluminous States* argue, shifts in sea ice, geopolitics, and economic landscapes are easing the extension of state sovereignty into volumetric space: "From the Arctic to the South China Sea, states are vying to secure sovereign rights over vast maritime stretches, undersea continental plates, shifting ice floes, and aerial volumes."

Yet for the communities and individuals who call the region home, the expectation of new infrastructures amid geopolitical proxy wars and melting sea ice is not about sovereignty as much as it is about everyday anxieties and anticipations. Seeking materialization through foreign investments and grassroots capital ventures, many of the proposed projects are abandoned, unbuilt, or sustained through perpetual promises. Nevertheless, the infrastructures' anticipation generates local economies and agencies.

Beyond the headlines and promotional materials of developers and investors, residents of coastal towns and potential inland transport hubs engage with visions of Arctic corridors to legitimize the construction and expansion of ports and railways, or conversely to resist "foreign" incursions into their lands and livelihoods. In other cases, the imagined, nascent, or latently operative infrastructures are used in unexpected ways, becoming sites of cultural and economic activity.

In this book, we follow diverse entanglements of the imagination with Arctic silk roads broadly conceived—from Scandinavia to Russia, Canada, and Alaska (see Map 0.1). This includes Arctic transport corridors directly envisioned as part of China's Polar Silk Road initiative but also broader evocations of trade with Asia and other distant markets through rapidly expanding and melting Arctic sea routes. Arctic silk roads exist as constellations of individual desires, agencies, and capital ventures that draw on a global imaginary to achieve social, political, and economic goals. We thus engage with the concept of silk roads in the plural heuristic sense to represent the imagination of global transport and exchange in the Arctic. Although, as we discuss in the following section, the concept originally emerged as a scholarly frame rather than emic articulation, many of the chapters in this volume show how in the 21st-century silk roads have taken on varying degrees of local resonance, in dialogue with their reification in official state policies, heritage development, and infrastructural planning. In addition to examining these emergent expressions, the concept allows us to think along the roads of our case studies to understand how the past is evoked in the present, interrogate how these imaginaries affect agencies, and contribute to broader questions in the anthropology of unrealized infrastructure.

Along the Russian Arctic coast, the national project of the Northern Sea Route (NSR) is circumscribed in state law and policy, carved by nuclear icebreakers controlled by the state company Rosatom (see Chapters 2 and 7 for historical discussion of the Northern Sea Route—from Soviet past to anticipatory geopolitical present). Following international sanctions with the country's invasion of Ukraine, the route has taken on new urgency as means for Russian economic self-sufficiency, and Chinese economic partnerships and "Polar Silk Road" initiatives continue to shape its development. Stammler, Ivanova, and Vitebsky in this volume, show that as state infrastructural revival and militarization along the Russian Arctic coast create expectations for energy, port, and other transport infrastructure that may or may not materialize, and many of the Chinese projects remain unrealized, they nevertheless generate a "charismatic agency" harnessed by local actors. This agency manifests as community projects with staying power beyond the horizons and ghosts of state infrastructural anticipation. Following the flurry of construction and renovation of community buildings in Tiksi, and the work of community leaders in Naybá who create jobs on the promise of shipping access via the NSR, the authors argue a point core to all chapters in this volume, that "the nature of infrastructure anticipations, and the possibilities for their fulfilment, are strongly connected with possibilities for local participation."

Across the chapters, evocations of "silk roads" in the Arctic or more general imaginings of global transport connectivity via northern shipping routes, are harnessed by local actors to enact desires for livelihood, community thriving, and wellbeing. The studies exhibit the multitemporality of infrastructural anticipation, wherein memories of development's ruins or yet-unrealized visions, come to bear on future anticipations. These "ghosts," evoked directly in the volume's chapters by Magnani and Magnani, and Stammler, Ivanova, and Vitebsky, suggest that Arctic silk roads, as is the case with many megaprojects, often manifest as *ghost infrastructures* conjured in the imagination rather than in their materiality, and that the infrastructures' physicality emerges instead through *economies of anticipation* generated by individuals and communities who live in the currents and shadows of the regional, state, and transnational visions. Amid recent geopolitical tensions, many of the infrastructural projects discussed in the book have been recast in the language of national security and promised state support that would replace reliance on foreign investment, yet they continue a "trajectory of anticipation" (*sensu* Stammler, Ivanova, and Vitebsky) that tie past and future infrastructural imaginings to serve people's day to day needs and desires.

The book begins in northeast China, from where imaginaries of "silk roads" emanate across the contemporary melting Arctic. Fraser shows how Ewenki reinforce China's geopolitical positioning as a "near-Arctic state" to develop cultural heritage institutions, livelihoods, and revitalization initiatives, as well as ethnic tourism and exchange with communities in Siberia and across the circumpolar North. The ethnography suggests that China's Polar Silk Road is as much cultural as it is infrastructural. Magnani and Magnani reveal similar cultural economic reverberations of infrastructural anticipation in northern Norway, where the promise and specter of a port's (unrealized) expansion to capture global markets is used instead to generate vibrant local economies from archaeological heritage to Sámi tourism, identity, and spirituality, which resurge in their physical entanglement with the port's infrastructure. Following these "economies of anticipation" enacted by local actors, the authors argue that the cultural and economic potentials of infrastructure are actualized in the anticipatory space "between operation and abandonment."

In addition to entrepreneurship, infrastructural anticipation can manifest through protest or ambivalence in relation to development. In Rovaniemi, known as the "capital of Finnish Lapland," the imagination of large-scale infrastructural networks resonates in everyday life through tourism. Amid expectations of increased connections for human movement in the "suspended world" trailing the COVID pandemic, Adams and Bennett (this volume) describe how local entrepreneurs innovated touristic infrastructures and activities in anticipation of the return of Chinese tourists, while Rovaniemi residents remained cautiously ambivalent or resistant. Lehtola's chapter, by following Finnish state dreams of an "Arctic Ocean Corridor" from 20th-century nation building to 21st-century visions of a "Silk Road" rail line across Sámi lands, points to an instance of cultural and environmental resistance as an unexpected materialization of otherwise mercurial infrastructural plans that have appeared and disappeared over the last century. The protest movement echoes emerging discussions in the region about the psychological pressures of ever-growing "planscapes" for resource extraction and development, which have been shown to exacerbate the cumulative impacts of past and existing industrial encroachments on Indigenous livelihoods, especially reindeer herding (Österlin and Raitio 2020). A range of resistance to ambivalence is similarly present in Nome, Alaska, where Povoroznyuk and Schweitzer describe plans for the "Arctic port of the US." Following over a century of "constant anticipation" surrounding new infrastructure and resources in Nome, local sentiments have steered away from enchantment (*sensu* Harvey and Knox 2016), toward a kind of silent refusal (*sensu* Simpson 2014), wherein concerns about Indigenous and other locally rooted marine resources and livelihoods, as well as housing and other services, are communicated through active disinterest. Amatulli finds a parallel trend of anticipatory entrepreneurship, ambivalence, and resistance in the ebbs and flows of infrastructural dreaming in Canada, where century and a half old plans of turning the port of Prince Rupert into a "gateway to Asia," have given rise to smaller, sustainable, and Indigenous-inspired food economies as expressions of opposition to continued port development, amid broader concerns about socioeconomic impacts.

Looking beyond the built environment toward the infrastructural imagination, the chapters collectively interrogate the social, political, and economic reverberations of the unbuilt. Whether people resist or harness development plans, infrastructural visions generate around them agentive relations. By examining envisioned projects through a multitemporal lens, we reveal contexts in which infrastructural

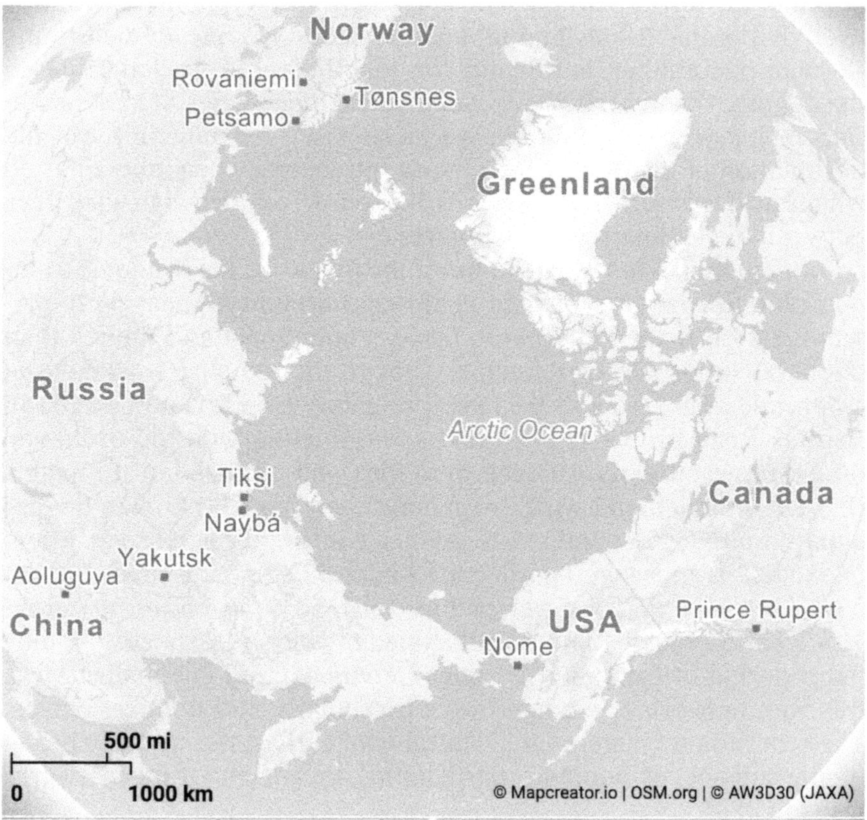

Map 0.1. *Map of key places in the book intersecting with visions of Arctic silk roads, 2025. Made by Natalia Magnani and Matthew Magnani. © Matthew Magnani and Natalia Magnani.*

imaginaries facilitate or constrain the agency of actors. We argue that the so-called "failed" ventures of global capitalism and its infrastructural promises materialize in unexpected ways through the local agencies and economies of anticipation that they generate. In other words, the unbuilt has material, social, political, and economic presence.

Silk Roads: The Revival of a Concept

As a concept, the "Silk Road" emerged in the 19th century to describe an ancient constellation of trade routes that spanned Europe and Asia, moving silk, glass, animals, and plants across vast distances from the last centuries BCE into the late-medieval period. While the origin of the term is not uncontested (Mertens 2019), the phrase was used by the mid-1800s and its popularization is typically tied to the German geographer, Ferdinand von Richthofen, in the latter half of the century (Waugh 2007). Archaeological discoveries of stray materials or ideas moving long distances show up for millennia, yet the designation of the Silk Road came to retroactively describe a period when an increased flow of goods moved across growing distances between expanding regional empires.

While popularized conceptions of the Silk Road may suggest a singular long-distance exchange coalescing around a linear path, elements of the trade routes were far less intentional and unified than might be suggested by its name. Though the regional movement of plants and animals spans well into prehistory (see Taylor et al. 2018), the Silk Road proper is imagined as originating with the exchange of silk between early Chinese dynasties and nomads on the plains (Liu 2010), a demand which expanded west past the Roman empire across multiple land and sea routes. Like any other extensive system of exchange, those who participated in trade along the roads did not have such an expansive perspective, instead contributing to countless local exchanges and barter systems. Only retrospectively, during a period of expanded European conquest and colonialism, was this vast network reimagined to suit scholarly and political climes. The rebranding bolstered the centrality of early states at the expense of the agency of nomadic participants in ancient trade networks (Misra 2020).

Despite its lack of cohesion as a concept in antiquity, silk roads, as a heuristic, have retained value as a scholarly term to frame the long-distance trade and intercultural exchange that took place

from around two millennia ago and endured through the medieval era (Knutson 2020; Franklin 2024). Diverse research, drawing on excavation, genetic evidence, and chemical analyses, have revealed patchworks of local agencies that, when approached at macro scale, illuminate long-distance exchange along major global corridors of trade. For instance, studies of mitochondrial genomes suggested intermixing between eastern and western populations in Central Asia during the Silk Road era (Comas et al. 1998)—a suggestion that has been explored in greater detail using autosomal data (Mezavilla et al. 2014). Isotopic analysis of glasswork hints at the long-distance movement of wares between the Middle East and Central Asia (Fiorentino et al 2019). Genetic analysis of other species, from horses to food crops, are equally suggestive of the broad range of sharing that occurred, often mediated not by large state actors on either side of the routes but by nomadic peoples who occupied the spaces between written records (Warmuth et al. 2013; Wang et al. 2022). Drawing on a range of ancient evidence, it is certain that across its multiple lengths, an even greater diversity of actors contributed to the movement and consumption of goods. These people formed social ties and shaped cartographies of exchange that would capture the attention of future publics and scholars.

Today, from Western Europe to the Middle East and East Asia, the route continues to be historicized and memorialized in text and public exhibitions that seek to foster local identities rooted in multicultural reciprocity (see Huerta 2021). UNESCO has added dozens of heritage sites along the Silk Roads, from Venice to Java, and programs bearing the same name buoy the peaceful international exchange of ideas (Shen et al. 2024). Scholars have emphasized the inherent risks of the persistence of the term, which may be used to reify divisions between East and West, or elevate national claims to sovereignty over borderlands (see again Franklin 2024; Knutson 2020). The ongoing revitalization of the concept represents a collision of complex ancient and contemporary socio-political landscapes, as a broad range of actors historicize interconnectivity in the past to encourage contemporary visions of local improvement. And just as in earlier centuries, the reimagination of the Silk Road has been co-opted by grassroots initiatives and regional actors as states develop high-profile international and future-oriented visions of trade and prosperity.

Established by European scholars as a construct to conjure new forms of contemporary exchange, past Silk Roads, as imagined, have taken on a life of their own. In 2018, China published a policy paper on the Arctic that brought plans for a "Polar Silk Road" to international consciousness (China's Arctic Policy 2018). The Road encourages a wide range of investment in infrastructure from ports to railways to energy tied to Arctic shipping routes, spanning the entirety of the circumpolar North and its navigable passages—Northeast, Northwest, and Transpolar.

China's Arctic goals articulate with its even more ambitious One Belt One Road strategy, also known as the Belt and Road Initiative, a push to develop and invest in infrastructures around the world such as roads, ports, railways, pipelines, and even digital technologies. The infrastructures are envisioned to form a broader network for trade and transport on land and sea, while evoking the connectivity and exchange of the historical Silk Road (Chan 2018). Just as the initial term "Silk Road" was developed retrospectively at the height of the colonial period to manifest new global ambitions, contemporary revivals of the term are embedded in broader geopolitical imaginaries that elevate intercultural trade and exchange to reach new ends, including governance (Winter 2021). At such a scale, it is not surprising that the Belt and Road is dotted with ghost infrastructures as much as operative constructions—many of the projects unfulfilled, abandoned, or built but suspended in latent or non-operational states.

Unrealized but idealized, memories of ancestral trade routes are being marshaled to support new geopolitical ambitions and unrestricted trade, connecting East and West through maritime and land routes. Conceptually, following "silk roads" allows us to analyze global networks past and present, not only as they materialize through built projects but also as they are imagined. Just as a discrete thoroughfare never existed historically, today a constellation of local agencies, aspiring toward the revitalization of global trade and interconnectivity, may be unified by current geopolitical ambitions to lay the groundwork for infrastructures underlaying a new silk road.

Arctic Passages in Context

If the evocation of ancient infrastructures conjures contemporary exchange, it is important to further situate the role of Arctic passages as long-term sites of human occupation and trade. Maritime highways through the Arctic have connected people for thousands of years; predecessors of contemporary Inuit communities from Greenland to Nunavut attest to the success of coastal adaptations. Their ancestors moved people and ideas across ice-covered corridors with kayaks and boats, subsisting from land and water and ice

(Friesen 2016). Coastal rock art depicting boating cultures appears in early Scandinavian rock art, while the northernmost regions of Russia have been settled with equally great time depth (Helskog 2014). Viking routes skirted Greenland and Iceland crossing the Atlantic, moving ivory and animals, and leaving durable expressions from Greenland to Newfoundland (Ingstad and Ingstad 2000). The Arctic waters and coastal resources that figure into contemporary imaginations have long facilitated movement and exchange.

Intersecting with these histories of settlement, the Arctic played an expanding role in the imagination of commerce and prosperity throughout the 16th century. In pursuit of improved trade routes, the region witnessed a dedication of resources by state actors and explorers endeavoring to economize travel between eastern and western ports and markets. Travel time around the Cape Horn in South America could be cut by the Northwest passage connecting the Atlantic and Pacific; the Northeast passage could allow ships to traverse Scandinavia across Siberian coastal waters and circumvent the African continent. The Scandinavian North Cape was rounded in the 1580s, and the Northeast passage traversed by the late 1870s (Armstrong 2011; Jacobsen 2015). The exploration of the Northwest passage, also beginning in the late 1500s, was hampered by ice even during the summer and was not crossed until 1906 (Williams 2010). These routes saw focused attention, but irregular crossing, due to thick and persistent sea ice through the late 19th and early 20th centuries.

A preoccupation with economizing trade reaches through the historical record to the present. Overlapping with expeditions to chart Arctic passageways in the 1800s and early 1900s, major infrastructural projects, including the Suez and Panama canals, carved pathways of water across land to circumvent entire continents. Massive rail systems were built to carry waves of raw materials, goods, and people. Despite the emergence of regular air transport and road systems since the middle of the last century, maritime trade remains essential to global commerce and energy production. Existing in perpetual states of anticipation and potential, the Northwest and Northeast passages stand to cut thousands of miles and nearly halve the distances and costs associated with major sea routes (Pharand 2007; Schøyen and Bråthen 2011). Both routes not only bypass greater distances translating to saved time and fuel, but also the geopolitics that come with reliance on narrow canal systems. Arctic passages remain more relevant than ever.

Building on the anticipation of shortcuts that reaches back centuries, travel across Arctic passages has increased in feasibility with rapid changes in global climate. Winter sea ice has reduced by over 3 percent every ten years in recent decades and nearly 70 percent of ice is now seasonal (Stroeve and Notz 2018; Roach et al. 2025). Full-year navigation of Arctic waters is predicted by 2100 (Zhao et al. 2024). Accelerated by climate change, questions of sovereignty, development, and security in the Arctic have continued to play out among a cadre of states with borders that cross the region. The United States, Russia, Denmark, Norway, and Canada, known as the "Arctic Five," control territories that jut physically into Arctic waters. The formalization of an Arctic Council in the 1990s further includes Sweden, Finland, and Iceland, launching from international discussions of security and scientific research that reach back to the 1980s (Koivurova and Vanderswaag 2007). Significant amounts of natural resources, including oil, natural gas, and battery metals, sit in Arctic soils and coastal waters. Investments in new technology to facilitate their removal and movement have become prominent, spanning well-established Russian initiatives to extract and transport liquid natural gas, to the exploration of the Scandinavian Arctic for battery minerals (Lavrenteva 2020; Dale and Gross 2023). The scope of projects, emergent and anticipated, would have been unimaginable just decades ago, and stand to resonate and collide with the visions of local actors across diverse cartographies.

With expanding stakes, Arctic shipping lanes attract the attention of nations with physical territory in the Arctic but also a growing list with Arctic-adjacent trade and security interests. Driven by interests in energy import and commerce, China's emergence as a "near-Arctic state" has accompanied the assertion of rights and governance without territorial sovereignty rooted in "scientific research, environmental protection, sea passages, and natural resources," positioning economic venture as moral principle through articulations of "harmony between natural conservation and social development" (China's Arctic Policy 2018). Along with this framing, the country's Arctic policy emphasizes sustainable development, climate change mitigation, international cooperation, and free navigation under existing international frameworks. However, these Arctic ambitions, including the development of a "Polar Silk Road," have been lacking in practical development in recent years despite the heightening of Russia–China cooperation, and the anticipation of more regular resource extraction and transport through the Northern Sea Route (Alexeeva and Lasserre 2018; Gunnarson and Moe 2021). With all eyes on the Arctic, a scramble for influence in the region has expanded beyond traditionally conceived forms of territorial sovereignty, following state security and commercial interests, scientific research, and tourism; these alternative expressions of governance have stoked anxieties anticipating the shape the region will take in the future.

Anthropology of the Unbuilt

Our focus on local anticipations of global infrastructures that exist in liminal states of physical realization, lends itself to intervention in the anthropology of the unbuilt. Social analyses of infrastructural plans and failed constructions have been taken up across disciplines, from human geography and history to anthropology. While many of these disciplines employ anthropology's trademark of ethnography, anthropological studies stand out in two main ways: first, the temporal dimension of fieldwork, which is often long term over multiple years and decades, thus facilitating deep ethnographic understanding of social change over time. Second, theoretical and methodological approaches in anthropology illuminate intersections between the micro and macro scales of social and material worlds—between everyday life, practice, and sociality, and national and global developments.

A burgeoning anthropological literature on the social lives of unrealized infrastructures examines the entanglement of people's daily strivings and imaginations with large-scale infrastructural development, politics, and bureaucracies, with research carried out over a period of months to years that allows for an analysis of change through time. For example, Harvey and Knox (2016) follow peoples' mundane experiences of road building over many years in Peru to suggest the role of enchanted affect in perpetuating infrastructures' promises. Similarly, Haines (2018) focuses on the day-to-day anticipations of a highway in Belize over a period of three years, to show how these anticipations shape Indigenous land and environmental politics.

Whether explicitly or indirectly, research on the not-yet-realized provides theoretical insight into experiences of time—the anticipation of what is to come or longing for that which could have been. It follows then that anthropological attempts to synthesize research on unfinished infrastructure have focused on material temporal dimensions—often as mediums through which politics and inequalities are reproduced (Anand, Gupta, and Appel 2018), or as heuristics to understand experiences of the unbuilt (Carse and Kneas 2019). An anthropology of unrealized infrastructure calls for the theoretical rethinking of temporality.

When it comes to growing anthropological interest in "megaprojects"—a term that may be used to describe diverse visions of silk roads in the Arctic, analyses of their distinguishing features have focused on transnational and economic scale, as well as decentralized qualities of implementation (Ullberg, Körling, and Strava 2023). However, it may also be argued that what distinguishes these projects is the scale of their unfinishedness, and thus their temporality.

Requiring coordination between a diversity of actors and interests at multiple scales, megaprojects are likely to emerge as patchworks of unrealized ghost infrastructures and the longing and anticipations for pasts and futures that accompany them (see Magnani and Magnani, and Stammler, Ivanova, and Vitebsky, this volume).

Projections of climate change and melting sea ice add yet another spatio-temporal dimension to unbuilt infrastructures as they materially and discursively reconfigure geopolitical relations. Scientific models predict shifts in seasonal ice cover and scales of environmental change but also new possibilities for shipping and technology—for example, what ice class for ships will be required to make the Arctic navigable and at what times of year. Meanwhile, infrastructures on land—ports, roads, railways, and industrial projects—are being planned in expectation of these changes. Both developments exist in anticipatory, abstract realms, tangibly manifesting in everyday life through the initiatives of individuals who act on the projections.

Methodologies for the Unbuilt

The study of infrastructure across disciplines follows networks of human, material, and/or technological relations. Anthropologists have focused especially on infrastructure's social, political, and economic entanglements (e.g., Harvey and Knox 2015). This is partly due to the tools with which anthropology is equipped. In his review essay in the early years of anthropology's infrastructural turn, Larkin (2013) noted the difficulty of examining "technological systems" using ethnographic methods. While scholars utilizing approaches in science and technology studies were generating a robust literature for the study of technological networks (e.g., Bennett 2010; Mitchell 2011), Larkin emphasized the equal importance of infrastructures beyond their technical qualities, namely as vehicles for "desire and fantasy." Such desires were subsequently taken up in studies of infrastructure's "promises" and "enchantments" (e.g., Harvey and Knox 2016; Anand, Gupta, and Appel 2018). Examining the promises of the unbuilt, this volume locates materiality not in the technical workings of structures but in dialogues between the imagination and social, economic, and political life. Although unrealized port, road, and railway networks cannot be physically engaged, their promises manifest materially through the agency of local actors who may resist or harness large-scale visions for on-the-ground ventures.

Examining infrastructural anticipation through historically rooted ethnography, the chapters do not exclusively follow the materiality of existing infrastructure but rather the relations of an infrastructural imagination of that which is not yet built. While some of the chapters also deal with the material entanglements and affordances of infrastructure (e.g., Povoroznyuk and Schweitzer, Magnani and Magnani), overall the case studies delve beyond an analysis of the built environment to ask what kinds of relations and agencies are engendered with the *imagination* of infrastructure, rather than its physical qualities. At the same time, we show how the imagination of infrastructure can manifest materially in ways that do not align with original blueprints—as local economies and ventures engendered by the infrastructural imagination. In doing so, we theorize an alternative materiality of infrastructure.

The authors of this volume examine infrastructure through a combination of ethnographic and historical methodologies. Each chapter follows myriad social and material relations surrounding infrastructures that are in various states of imagination, construction, or abandonment. The authors of the volume are social and cultural anthropologists, human geographers, and historians, with expertise across other disciplines and sub-disciplines such as archaeology, legal, and Indigenous studies. They have long-term relationships with the people and communities of what has been conceived as the circumpolar North–from Sápmi to Siberia, Alaska and Canada, and the self-proclaimed "near-Arctic" China from which a "Polar Silk Road" emanates. Their ethnographic perspectives on the anticipatory and imaginative aspects of infrastructure, instead of their physicality, makes the primary focus of the book more on the social rather than technological dimensions of infrastructure. As we follow the social, political, and economic relations engendered through the imagination of infrastructure, we also ground the work in a broader historical and geopolitical context: Amatulli, Lehtola, and Povoroznyuk and Schweitzer connect long-term and nationally driven infrastructural anticipations with present-day ambivalences or resistance against development in Canada, Finnish Sápmi, and Alaska (respectively), while Stammler, Ivanova, and Vitebsky follow "trajectories of anticipation" from Soviet policies and ideologies of uncurbed expansion to new mobilizations of infrastructural dreams in the Siberian Arctic.

Existing in the imagination as visions or ruins, infrastructures are not yet, or no longer, "matter that enables the movement of other matter" (Larkin 2013). Instead, they comprise non-matter that engenders matter. Our multitemporal ethnographic methodologies allow us to examine this translation.

Following Roads

The everyday lives of people in the circumpolar North and "near-Arctic" nations geopolitically tethered to the region, are entangled with local political bodies, national agendas, and international agreements that collectively govern the Arctic. Visions of global transport and connectivity across these areas are variably experienced as top-down blueprints and development ventures, or on-the-ground actions of resistance, heritage making, and entrepreneurship.

Drawing together the ethnographic experiences of this volume, we suggest that the degree to which grand infrastructural visions are engaged locally—either through their embrace or resistance—depends on the degree to which they intersect with everyday lives and livelihoods. Where the anticipation of the not-yet-realized, or its preliminary infrastructures, may be harnessed to generate local economies, anticipation takes on alternative material forms. Many of the book's chapters focus on cases of tourism, community, and heritage infrastructures that co-opt or anticipate geopolitically strategic visions of development (Adams and Bennett; Fraser; Magnani and Magnani; Stammler, Ivanova, and Vitebsky). As Stammler, Ivanova, and Vitebsky emphasize, the degree to which large-scale anticipations are actualized as local agencies depends on questions of scale and participation.

Following this logic, projects that are distant from community and economic realities are rejected or barely register in everyday discourse. Povoroznyuk and Schweitzer were surprised to find that a large port project in Nome, Alaska, was greeted with disinterest by a majority of local residents, with questions raised instead about practical matters such as housing—concerns also prominent surrounding port plans in Prince Rupert discussed by Amatulli. Resistance to port development was present to some degree in both cases, but numbed by general apathy (Povoroznyuk and Schweitzer), or manifesting indirectly as entrepreneurial forms of subsistence production (Amatulli).

Similarly, state dreams for roads and railways across Sámi lands to the Arctic Ocean, discussed by Lehtola, have been met with mixtures of ambivalent resignation and protest through the last century.

The chapters follow visions of connectivity across the Arctic to reveal potential for local agencies surrounding large-scale infrastructural projects that may or may not be realized. When people envision the use of infrastructure or its promises for their own livelihoods (i.e., tourism, political positions, investment opportunities), or conversely see them as endangering their subsistence or other lifeways (i.e., land-based livelihoods or businesses in the vicinity of expanding infrastructure), their imaginations and actions become entangled with the threats or possibilities of development. When national planning and regional developers' dreams are top-down and not directly tied to peoples' everyday economies of well-being, social networks, or political aspirations, they are less likely to draw action or attention.

In these ways, we show that global capitalism only "fails" when the projects remain at large and theoretical scales that do not intersect with day-to-day lives. In most cases, state and international visions for connectivity via expansions of infrastructure and commerce are harnessed at smaller scales long before the laying of cement and iron, as local entrepreneurs and activists act on these blueprints to achieve their own social, economic, and political goals.

Natalia Magnani is Assistant Professor at the Department of Anthropology and Climate Change Institute at the University of Maine. She leads the Norwegian Research Council project "Arctic Silk Road" based at UiT The Arctic University of Norway, whose participants have collaborated in shaping this volume. She has published numerous articles on the Sámi regions, including in *Current Anthropology* and *American Anthropologist*, and has two books forthcoming—*Silent Knowledge: Reviving Sámi Practice* (University of British Columbia Press) and *The Craft of Belonging: Material Culture and Social Boundaries in Sápmi* with Matthew Magnani (University of Toronto Press).

Matthew Magnani is Assistant Professor of Anthropology and Climate Change at the University of Maine, and a researcher on the "Arctic Silk Road" project based at UiT The Arctic University of Norway. He has published dozens of articles on material culture past and present, in journals such as *American Antiquity* and *Current Anthropology,* and is the co-author of *The Craft of Belonging: Material Culture and Social Boundaries in Sápmi* with Natalia Magnani (University of Toronto Press).

References

Alexeeva, Olga, and Frederic Lasserre. 2018. "An Analysis on Sino Russian Cooperation in the Arctic in the BRI era." *Advances in Polar Science* 29(4): 269–82.

Anand, Nikhil, Akhil Gupta, and Hannah Appel, eds. 2018. *The Promise of Infrastructure*. Durham: Duke University Press.

Armstrong, Terence. 2011. *The Northern Sea Route: Soviet Exploitation of the Northeast Passage*. Cambridge: Cambridge University Press.

Bennett, Jane. 2010. *Vibrant Matter: A Political Ecology of Things*. Durham: Duke University Press.

Billé, Franck. 2020. *Voluminous States: Sovereignty, Materiality, and Territorial Imagination*. Durham: Duke University Press.

Carse, Ashley, and David Kneas. 2019. "Unbuilt and Unfinished: The Temporalities of Infrastructure." *Environment and Society* 10(1): 9–28.

Chan, Man Hung Thomas. 2018. "The Belt and Road Initiative—the New Silk Road: A Research Agenda." *Journal of Contemporary East Asia Studies* 7(2): 104–23.

"China's Arctic Policy." 2018 (January 26). Beijing: The State Council Information Office of the People's Republic of China. Retrieved March 2025 from https://english.www.gov.cn/archive/white_paper/2018/01/26/content_281476026660336.htm.

Comas, David, Francesc Calafell, Eva Mateu, Anna Pérez-Lezaun, Elena Bosch, Rosa Martínez-Arias, Jordi Clarimon, Fiorenzo Facchini, Giovanni Fiori, Donata Luiselli, Davide Pettener, and Jaume Bertranpetit. 1998. "Trading genes along the silk road: mtDNA sequences and the origin of central Asian populations." *The American Journal of Human Genetics* 63(6): 1824–38.

Dale, Ragnhild Freng, and Lena Gross. 2023. "The Arctic: Last Frontier for Energy and Mineral Exploitation?" In *Handbook on International Development and the Environment*, ed. Benedicte Bull and Mariel Aguilar-Støen, 154–69. Cheltenham: Edward Elgar Publishing.

Fiorentino, Sara, Barbara Venezia, Nadine Schibille, and Mariangela Vandini. 2019. "Streams Across the Silk Roads? The Case of Islamic Glass from Ghazni." *Journal of Archaeological Science: Reports* 25: 153–70.

Franklin, Kate. 2024. "Archaeology of the Silk Road: Challenges of Scale and Storytelling." *Journal of Archaeological Research* 32: 263–308.

Friesen, Max. 2016. "Pan-Arctic Population Movements: The Early Paleo-Inuit and Thule Inuit Migrations." In *The Oxford Handbook of the Prehistoric Arctic*, ed. Max Friesen and Owen Mason, 673–92. Oxford: Oxford University Press.

Haines, Sophie. 2018. "Imagining the Highway: Anticipating Infrastructural and Environmental Change in Belize." *Ethnos* 83(2): 392–413.

Jacobsen, Jens Kr. Steen. 2015. "North Cape: In the Land of the Midnight Sun." In *The Tourism Imaginary and Pilgrimages to the Edges of the World*, ed. Nieves Herrero and Sharon R. Roseman, 120–40. Bristol, Blue Ridge Summit: Channel View Publications.

Gunnarsson, Björn, and Arild Moe. 2021. "Ten Years of International Shipping on the Northern Sea Route." *Arctic Review on Law and Politics* 12: 4–30.

Harvey, Penny, and Hannah Knox. 2016. "The Enchantments of Infrastructure." In *Roads and Anthropology*, ed. Dimitris Dalakoglou and Penelope Harvey, 63–78. London: Routledge.

Harvey, Penny, and Hannah Knox. 2015. *Roads: An Anthropology of Infrastructure and Expertise.* Ithaca: Cornell University Press.

Helskog, Knut. 2014. *Communicating with the World of Beings: The World Heritage Rock Art Sites in Alta, Arctic Norway*. Oxford: Oxbow Books.

Huerta, Ricard. 2021. "Silk Road Museums: Design of Inclusive Heritage and Cross-Cultural Education." *Sustainability* 13(11): 6020.

Ingstad, Helge, and Anne Stine Ingstad. 2000. *The Viking Discovery of America: The Excavation of a Norse Settlement in L'Anse aux Meadows, Newfoundland.* St. John's: Breakwater Books.

Knutson, Sara Ann. 2020. "Archaeology and the Silk Road Model." *World Archaeology* 52(4): 619–38.

Koivurova, Timo, and David L. Vanderzwaag. 2007. "The Arctic Council at 10 Years: Retrospect and Prospects." *UBC Law Review* 40: 121.

Larkin, Brian. 2013. "The Politics and Poetics of Infrastructure." *Annual Review of Anthropology* 42(1): 327–43.

Lavrenteva, Nina. 2020. "Energy Policy in the Arctic: Yamal LNG in Russian International and Domestic Political Agenda." *The Journal of Cross-Regional Dialogues/La Revue de dialogues inter-régionaux* 1: 83–104.

Liu, Xinru. 2010. *The Silk Road in World History*. Oxford: Oxford University Press.

Mishra, Ravi K. 2020. "The 'Silk Road': Historical Perspectives and Modern Constructions." *Indian Historical Review* 47(1): 21–39.

Mertens, Matthias. 2019. "Did Richthofen Really Coin 'the Silk Road'?" *The Silk Road* 17(1): 1–9.

Mezzavilla, Massimo, Diego Vozzi, Nicola Pirastu, Giorgia Girotto, Pio d'Adamo, Paolo Gasparini, and Vincenza Colonna. 2014. "Genetic Landscape of Populations along the Silk Road: Admixture and Migration Patterns." *BMC Genetics* 15: 1–10.

Mitchell, Timothy P. 2011. *Carbon Democracy: Political Power in the Age of Oil.* New York: Verso.

Österlin, Carl, and Kaisa Raitio. 2020. "Fragmented Landscapes and Planscapes: The Double Pressure of Increasing Natural Resource Exploitation on Indigenous Sámi Lands in Northern Sweden." *Resources* 9(9): 104.

Pharand, Donat. 2007. "The Arctic Waters and the Northwest Passage: A Final Revisit." *Ocean Development & International Law* 38(1–2): 3–69.

Roach, Lettie A., Madison M. Smith, Agnieszka Herman, and Damien Ringeisen. 2025. "Physics of the Seasonal Sea Ice Zone." *Annual Review of Marine Science* 17: 355–79.

Schøyen, Halvor, and Svein Bråthen. 2011. "The Northern Sea Route versus the Suez Canal: cases from bulk shipping." *Journal of Transport Geography* 19(4): 977–83.

Shen, Yingning, Junmin Liu, Jianan Han, and Xiang Wan. 2024. "Reconstructing the Silk Road Network: Insights from Spatiotemporal Patterning of UNESCO World Heritage Sites." *Land* 13(9): 1401.

Simpson, Audra. 2014. *Mohawk Interruptus: Political Life Across the Borders of Settler States.* Durham: Duke University Press.

Stroeve, Julienne, and Dirk Notz. 2018. "Changing State of Arctic Sea Ice Across all Seasons." *Environmental Research Letters* 13(10): 103001.

Taylor, William, et al. 2018. "Early Pastoral Economies along the Ancient Silk Road: Biomolecular Evidence from the Alay Valley, Kyrgyzstan." *PloS One* 13 (10): e0205646.

Ullberg, Susann Baez, Gabriella Körling, and Cristiana Strava. 2023. "Making Megaprojects: The Practices and Politics of Scale-Making." *Ethnos* 90(2): 1–10.

Wang, Guangyan, Qian Chen, Ya Yang, Yuanwen Duan, and Yongping Yang. 2022. "Exchanges of Economic Plants along the Land Silk Road." *BMC Plant Biology* 22(1): 619.

Waugh, Daniel C. 2007. "Richthofen's 'Silk Roads': Toward the Archaeology of a Concept." *The Silk Road* 5(1): 1–10.

Williams, Glyn. 2010. *Arctic Labyrinth: The Quest for the Northwest Passage.* Berkeley: University of California Press.

Winter, Tim. 2021. "Geocultural power: China's Belt and Road Initiative." *Geopolitics* 26 (5): 1376–99.

Zhao, Pengjun, Yunlin Li, and Yu Zhang. 2024. "Ships Are Projected to Navigate Whole Year-Round Along the North Sea Route by 2100." *Communications Earth & Environment* 5(1): 407.

CHAPTER 1

Imagining the Arctic in Northeast China

Ethnic Minorities and the State along the Polar Silk Road

Richard Fraser

China consists of fifty-six officially recognized ethnic groups: the majority Han, who make up 91 percent of the population, and fifty-five so-called "ethnic minorities" or "ethnic nationalities" (*shǎoshù mínzú*), who number the not insignificant amount of 110 million people. Since the founding of the People's Republic (PRC) in 1949, the Chinese state has actively sought to nativize its ethnic minorities, specifically by downplaying their international origins, their connections to neighboring or cross-border groups, and strategically managing their "non-Chinese" characteristics (e.g., religion). Thus, despite officially being a multi-ethnic state—and with all ethnic minorities afforded cultural protection in the PRC Constitution—there has always been a strong unifying trend in China's ethno-political project (Bulag 2010). This was already visible during the early years of the PRC, when several major ethnic groups came together to form a "United Front" against Japanese occupation—something still valorized in contemporary political discourse. And this is similarly seen today in public and academic discussions; for example, sociologist Ma Rong (2007) has advocated a depoliticized approach to ethnicity that emphasizes economic development and social integration over identity politics. Opposed to the "melting pot" thesis of many

Western societies, Ma argues against ethnic differentiation that he believes reinforce divisions and instead advocates a unified national identity. Another apt example is the reclamation of Mongol cultural heroes by Chinese historians (e.g. Chinggis Khan in Ordos), while in several border regions minority ethnonyms and places have been re-ascribed with Chinese names to better emphasize a Sino-centric version of history.

This is something similarly seen in the case of Ewenki. For example, during the Qing Dynasty the Manchu rulers established the Eight Banners System (*Baqi* 八旗), which sought to integrate the northern groups living in the region (Elliot 2001, Mullaney 2011). They included Daur, Orochen, and Ewenki, who the Qing authorities integrated under the label "Solon" (*Suolun*), which was an ethnic-political categorization aimed at facilitating incorporation into the Manchu state. As Loretta Kim points out (2009: 18), "two or more groups that considered themselves to be distinct could become a single group for bureaucratic purposes such as census taking and collection of taxes or tribute" (see also Dumont 2017). This nativization continued even after 1949, when the state classified Ewenki using the Chinese ethnonym *Ewenke*, specifically to differentiate them from Evenki and Eveny groups across the border in Siberia—i.e. firmly establishing them as "*Chinese* Ewenki."

In recent years, however, an interesting development has occurred that runs counter to the nativization process, namely, a strengthening of ties between Ewenki and their cross-cultural cousins in Siberia, as well as new connections—both real and imagined—being forged with other Indigenous peoples in the Circumpolar North. This has been actively supported by the national state and Ewenki minority government, through funding and complemented by a wider academic–museological recognition that Ewenki are—in fact—subarctic Indigenous peoples with their roots in Siberia. Coupled with this, the area they inhabit—the *Da Xingangliang* Mountains—has been strategically re-defined as a subarctic zone with close cultural and ecological ties to the Arctic. This is best encapsulated in contemporary tourism narratives and policies which describe the region as "Chinese Siberia" (*Zhōngguó xībólìyǎ*).

All of this is a highly complex process that is taking several forms—only some of which I will address in this chapter. The first is culturally, through the support and promotion of reindeer herding and cultural heritage amongst Ewenki and other northern minorities, including cultural and linguistic revitalization, and the expansion of an ethnic tourism industry. Second is environmentally, through scientific research and climate change mitigation across the Sino-Russian border, including conservation initiatives with highly symbolic Arctic species such as the Siberian tiger. And third is infrastructurally, through cross-border trade with Russia, increasing ties with other Arctic states, and the building of infrastructure, transportation networks, and the expansion of the resource industries—including the Polar Silk Road. All of this is having the dual effect of increasing connections between China, Russia, and other Arctic states while simultaneously building the narrative that China is—and will be—a prominent player in Arctic affairs.

In this chapter I focus primarily on the cultural dimensions of this process, though I will also touch upon both infrastructure and development. And what I argue is that these threads are all tying together to form part of a broader re-imagining of China as a "near Arctic state"—a term that has garnered increasing attention in recent years, underpinning as it does China's present and future engagement in Arctic affairs—and, of course, forms the ideological basis for the Polar Silk Road (see Introduction). In practice, these threads are fragmented: they intersect and influence one another in various ways, and they include both implicit and explicit dimensions. So, on the one hand, there are direct government policies, allocations of funding, academic and scientific research, and the physical building of infrastructure which seeks to increase China's role in the Arctic. At the same time, there are more implicit ideas surrounding culture, history, and heritage, which similarly shape narratives and imaginings. This can be seen in a short excerpt from the Arctic Institute, an independent think tank based in Washington, which describes China's Arctic connections in this way:

> China's northern roots stretch back farther than is generally known. In 1644, the Manchus invaded Beijing from the north, establishing the Qing Dynasty. The Manchus are a Tungusic people whose ancestral homeland lies in Siberia, and their language falls within the same family as Even and Ewenki, spoken by the Siberian reindeer-herding peoples of the same names. Some Evenks, who are generally recognized as Arctic Indigenous people, still live in pockets of northern China, where some still practice reindeer herding. (The Arctic Institute 2022)

Here we can see a very clear connection being drawn between China's Arctic roots and the cultures and histories of its northern ethnic minorities—as well as an assumption that China has always had a much deeper tie to the Arctic than is generally assumed. Of course, I am not suggesting that the Chinese state explicitly uses Ewenki and other northern minorities as part of their legal claim for the country's "near Arctic status." But what I am suggesting is that in the process of China pivoting to the Arctic—and imagining itself as a "near-Arctic state"—a new space has opened which simultaneously co-opts—but also benefits—Ewenki in several ways. And what I seek to explore is how Ewenki (and other minorities and Indigenous people) become *interpellated*—in Louis Althusser's (2014) sense of the term—into wider state projects and imaginings, and how we might conceptualize and understand this dual process of co-option and benefit—of interpellation and agency—as minority and Indigenous people are integrated, enveloped, and "put to work" in state projects such as the Polar Silk Road.

To put this in simple terms: I argue that given China's pivot to the Arctic it is symbolically and materially useful for the state to have (and give support to) a community of subarctic reindeer herders within its borders; that Ewenki have been admitted into pan-Arctic organizations such as the Association of World Reindeer Herders; and that the country includes a large part of its territory officially classified as subarctic taiga (including having iconic species such as the Siberian tiger, the Arctic fox, and the opportunity to see the Aurora Borealis). And, as I will show below, this has both positive and negative consequences for Ewenki themselves. For example, it has resulted in unprecedented levels of support for cultural and linguistic revitalization, various forms of cross-cultural exchange, as well as the allocation of substantial levels of funding for settlement re-development, museum construction and expansion, and new opportunities within the ethnic tourism industry

So here I see a kind of feedback loop: Ewenki both benefit from, but are also drivers of, what Tim Winter (2021) calls a geo-cultural imaginary—in this case, the imaginary of China as a near-Arctic state. And here it is important to emphasize that the term *imaginary* does not imply a superficial or merely constructivist idea without power or consequence. As Cornelius Castoriadis (1987) describes, a cultural imaginary is the collective set of shared meanings, symbols, and representations that shape and give coherence to a society's worldview, which is actively created and constantly reimagined by its members through social practices and institutions. Indeed, as I will show through some ethnographic examples, imagining China as an Arctic state carries real weight, not only in the form of investment, infrastructural development, and resource extraction, but also, in terms of raising hopes and ambitions—as well as fueling fears and apprehensions—in the minds of those who might be affected. Seen from this perspective, notions of culture, history, and heritage are immensely effective forms of soft power, in shaping perceptions and narratives about China's role in the Arctic—and the Polar Silk Road more generally—both domestically and around the world.

The Aoluguya-Ewenki in Northeast China

Officially, Ewenki in China consist of 30,000 people, but this is complicated given there are three Ewenki subgroups, the smallest of which is the Aoluguya Ewenki, who consist of just 270 people. They are China's only reindeer-herding community who traditionally were separated into different clans associated with specific hunting grounds (Lindgren 1933; Nentwig 2003). In the past, they crossed back and forth over the Amur-Heilongjiang River relatively freely and, although some had interactions with the Qing state during the early 20th century, others only entered Chinese territory as late as the 1920s. Still today, there are elderly generation Ewenki who can speak some Russian, have Russian names, and have been influenced by Russian Orthodox Christianity.

Culturally, Ewenki represent an apt example of what Hugh Beach and Florian Stammler (2006) call "symbiotic domestication"—that is, they are primarily hunters who use their reindeer as pack animals and for subsistence milking. Traditionally, they lived in tepee-like tents and moved frequently between hunting grounds, which necessitated intimate relationships to animals and the taiga and resulted in both animist ontological experiences of the environment and of nonhuman personhood—what David Anderson (2000) calls a "sentient ecology."

With the founding of the PRC in 1949 the state embarked upon a sedentarization campaign and built the first Ewenki settlement close to the Russian border—called Aoluguya. This was accompanied by increased state support in the form of healthcare and education, leading to the raising of life expectancy. Despite sedentarization, however, mobility and hunting remained central aspects of everyday life for the next few decades.

From the 1980s, Ewenki saw an increase in support as the state emphasized modernization, including upgraded housing in Aoluguya and various affirmative action policies such as welfare payments and exemption from the one-child policy. Also at this time, they experienced the marketisation of the reindeer economy, with antlers cropped every year and processed for sale in the form of Traditional Chinese Medicine (TCM) (Beach 2003; Fraser 2010).

By far the most significant change, however, was in 1998 when the state banned hunting. This was justified on the grounds of environmental protection and the supposed reduction in wild animals because of over-hunting (Xinhua 2003). In fact, it was more due to the expanding timber industry and the mass immigration of Han Chinese workers (Lundberg and Zhou 2009). Today, people still feel the lingering effects of the hunting ban, which has led to increased rates of alcoholism, suicide, and social dislocation. And of course, all of this has been exacerbated by increased cultural assimilation, generational changes, and language-loss.

In 2003, when I arrived to do my first fieldwork, Ewenki experienced another major change in the form of a forced relocation. Here the entire community—and their 1,200 reindeer—were moved 250 km south to a new purpose-built settlement on the outskirts of Genhe City, also called Aoluguya. This was justified as a case of "ecological migration" (*shengtai yimin*), which is a specific type of resettlement in China involving the movement or sedentarization of semi-nomadic or nomadic herders and hunters due to perceived environmental degradation (Yeh 2005; Kolas 2011). In fact, the relocation was more about the state development of a peripheral ethnic minority people and the expansion of the wider area for infrastructure and development (Fraser 2021).

It is significant to note that the Ewenki homeland is a center for oil and natural gas extraction, and this is where the Polar Silk Road first comes into play. Situated nearby is the Daqing oil field, the largest oil field in China. Opened in the 1980s, it was expanded around the time of the relocation and is set to develop further as a new oil reserve was discovered in 2021. The area is also home to significant natural gas pipelines which import gas from Russia. The first of these is the main arm of the *Power of Siberia* pipeline—which has an extension running through northeast China from Blagoveschensk-Heihe all the way to Shanghai. It is called the Heihe-Shanghai pipeline and has a tunnel under the Amur-Heilongjiang River close to the original Aoluguya township. The second is the Chinese Russian Oil Pipeline (CROP) from Mohe to Daqing, which also traverses the Da Xingangliang mountains and through areas previously used by Ewenki as reindeer pastures.

Seen from this perspective, the relocation was fundamentally tied to the resource industries and the wider infrastructural development of the region. Looking ahead, the area is set to become an important node along the Polar Silk Road, as new pipelines have been proposed across the border to southern Siberia and gas is already being shipped from the region along the Northern Sea Route. In fact, you can follow the movement of these resources from northeast China all the way to the Yamal Peninsula, as well as down again through pipelines running the length and width of Russia.

The 2003 relocation was considered by many as a failure. And that is because the state had sought to domesticate the Ewenki's reindeer by constructing pens inside the new settlement. But reindeer are semi-wild animals dependent on wild lichen and so only a few months after relocation the reindeer became weak and some died, leading to a dramatic conflict where the herders broke down the pens and returned to the forest with their animals. Following this, the state compromised and allowed Ewenki to maintain their campsites in the forest, all the while seeking new opportunities for economic development.

Today, life for Ewenki exists "in-between the city and the taiga." Some people remain permanently in the settlement and work within the growing tourism industry; others move back and forth between the settlement and the campsites; and a few spend virtually all their time in the forest as herders. As of 2024, there were twelve campsites, some just 5 km from the new Aoluguya and others hundreds of kilometers away close to the Russian border. There were approximately 1,200 reindeer, with herders practicing a limited form of mobile reindeer pastoralism. Through this, they acquire and sell reindeer antler and other reindeer products (placenta, organs) aimed at supplying the TCM market, which allows for a steady, though limited, income. In recent years, tourism has become a growing sector, with visitors drawn to Ewenki cultural practices and landscapes, which some herders leverage by offering guided tours or tent stays in the forest. Craft production, especially items made from birch bark, antlers, and forest materials, provides an additional source of income.

While this has created new opportunities, the situation in the campsites—and for the reindeer—remains complicated. The herders are subject to increasingly stringent measures, including the restriction of reindeer herding in conservation areas and harsher penalties for harvesting forest resources and lighting fires during certain seasons (Fraser 2021). The situation is made more complicated by the fact that only a handful of younger generation Ewenki want to live and work in the taiga, though recent investments in solar panels, mobile phone technology, and Wi-Fi have started to address this. Meanwhile, hunting continues to be banned despite increasing numbers of predators. Every year, the herders lose dozens of reindeer because of animal attacks, as well as from illegal poaching, which remains the major source of contention with the local state.

Constructing "Chinese Siberia"

In northeast China, the emerging tourism industry and its accompanying infrastructural development have taken on a distinctive form, shaped by the region's unique environmental conditions. The taiga forests, extreme cold, long snowy winters, and subarctic wildlife—alongside the ethnic minorities Indigenous to the area—have all played a central role in this transformation.

The region's tourism narrative is deeply rooted in the cultural–ecological relationship between the Daxingangling Mountains and the so-called "forest-based," "subarctic," and "hunting and reindeer cultures" of the Ewenki and neighboring Orochen. Within this framework, the area is often marketed as "Chinese Siberia," a branding reflected in its tourism infrastructure. Attractions and destinations carry evocative names such as "The Arctic Village," "The Coldest City in China," and "The City of the Northern Lights" (Fraser 2020). This branding appeals to tourists seeking outdoor, eco-based activities like hiking, snowmobiling, and camping, as well as those interested in experiencing ethnic minority cultures. Orochen are frequently portrayed as "China's Last Hunter-Gatherers," while Ewenki are highlighted as the country's "only reindeer herders."

This has led to a rapidly expanding ethnic tourism market, attracting visitors interested in purchasing traditional handicrafts, foraged forest products like berries and mushrooms, and hunted animal pelts and organs valued in TCM (Fraser 2020; Fraser forthcoming). When I first visited the region in 2008, ethnic tourism was almost nonexistent. However, over the past decade, substantial investments totaling billions of RMB have dramatically transformed the landscape. Across the northeast, minority villages, museums, and cultural heritage parks are being built or renovated. In Ewenki and Orochen regions, architectural styles are being reimagined to incorporate ethnic motifs, while broader rural development initiatives prioritize housing upgrades, road expansion, improved connectivity, and the creation of protected areas for eco-tourism and conservation (Fraser 2022).

This shift is evident in the case of the Aoluguya Ewenki Museum, which was upgraded in 2018. Built in the 1980s, it was typical of the time's approach to museums in minority regions, consisting of several sparsely decorated rooms with diorama-style displays of "traditional" Ewenki life. These included poorly crafted figurines of hunters, replicas of traditional tents, and numerous stuffed animals surrounded by plastic foliage.[1]

The museum also featured enlarged historical photographs depicting Ewenki life. These images focused prominently on state health care and education campaigns, as well as the supposed benefits of sedentarization. Information panels were attached to the photographs detailing Ewenki history, from the Qing Dynasty to the establishment of the PRC in 1949. The panels followed a clear unilinear and nativist narrative, depicting Ewenki as moving from a "primitive tribal" society to an advanced form of "Chinese socialism" with the help of the communist state, as well as reaffirming their place in the new Chinese nation.

By 2018 the situation had changed considerably. Here the state opened a new, highly sophisticated, and ambitious museum, blending cutting-edge design with a commitment to the heritage of Ewenki culture. Housed in a sleek, contemporary building in the center of town, visitors explore immersive exhibits featuring a range of artifacts, from hunting tools to intricate textiles. Interactive digital displays allow one to engage with the culture through multimedia presentations, including virtual tours of Ewenki camps and detailed timelines of the community's history.

The museum is also a hub for Ewenki art, showcasing traditional paintings and some newer works of modern Ewenki artists. One of the most striking features is the high-tech exhibition spaces, which bring ancient stories to life, allowing visitors to interact with historical narratives and cultural legends in an immersive environment. In addition to its rich cultural exhibitions, the museum includes a state-of-the-art auditorium where visitors can attend lectures, screenings, and performances. There are also dedicated spaces for workshops and educational programs, where heritage transmitters teach visitors traditional crafts and offer hands-on experiences of Ewenki craftsmanship. The museum café and gift shop offer a selection of locally crafted goods, books, and unique cultural products, further enhancing the connection between the past and present. Yet the museum is just one example of the wider infrastructural changes to occur across the region. In the section below, I look more closely at how Ewenki have become caught up in this transformation, but also, how they themselves are benefitting from the new opportunities associated with China's pivot to the Arctic.

The Aoluguya Ethnic Reindeer Resort

In 2012 Aoluguya underwent a major refurbishment, upgraded and integrated into the cultural heritage (*wenhua yichan*) and ethnic tourism industries. This is typical of China's minority regions where tourism and heritage are seen as a viable route for rural economic development, as well as for the so-called 'protection' of minority culture (Oakes 2005, 2013). Here communities are offered modernized housing and improved facilities in what are often relatively poor and/or remote regions, while the state creates alternate livelihood opportunities through the development of an ethnic tourism market.

What was most interesting about the 2012 refurbishment is that the entire settlement was modelled on, and inspired by, Sámi architecture in Sápmi.

So how did this come about? First in 2007, members of the Ewenki minority government and several Ewenki researchers visited Karasjok—one of the main Sámi settlements and location of the Sámi parliament in Norway. This was funded by the national government and partly organized by the China-Nordic Arctic Research Centre in Shanghai. Later in 2008, representatives of the Association of World Reindeer Herders (specifically Johan Mathis Turi and Elna Sara) visited Aoluguya—where they attended the so-called "Aoluguya reindeer herder and minority culture festival." They participated in a research seminar and held meetings with scientists and administrators in the region. This achieved the goals of the visitors learning more about reindeer herding in China but also for Ewenki to gain knowledge and experience from their Sámi counterparts.

Second, and building upon this, the Ewenki minority government hired a Finnish consultancy firm (Pöyry) in 2011 to oversee the re-development of Aoluguya. They visited Karasjok and Kautokeino and, in 2012, went on to re-design Aoluguya as a tourism-eco-settlement: officially called the "Aoluguya Ethnic Reindeer Resort." It is interesting to consider for a moment all of the cultural flows and imaginings behind this: here we have an ethnic minority in northeast China, consisting of just 270 people, with their roots in Siberia, having their hometown re-designed by a Finnish consultancy firm, based on an idea of Sámi reindeer herding culture, all of which was funded by the Chinese state.

The re-development was not insignificant. The total project cost 100 million RMB—which is substantial for a community of just 270 people. Today, the new Aoluguya includes housing for all Ewenki families, a 150-room hotel, a restaurant, and a new state-of-the-art museum. The entire settlement is made from high-quality Russian-imported pine. There is an industrialized antler-processing and packaging factory, and a handicraft center where elders and heritage transmitters manufacture traditional crafts—as well as a shop for visiting tourists. Just outside of town there is a reindeer camp for tourists to eat, drink, and take photographs. And, of course, there is wider supportive infrastructure in the form of roads, lighting, walkways, etc.

It is also important to note that all of this was paid for by the state. Each family received a house for free; they do not pay for electricity, water, or heating; and they receive subsidized health care, schooling, and elderly care. There are buses transporting people from Aoluguya to the nearest city (Genhe), and jeeps transporting the herders back and forth between the campsites and the settlement. But of course, the new Aoluguya has a dual function: on the one hand, it provides housing for the community, but also, to present a tourism experience. Indeed, Aoluguya is officially described (in the tourism literature and online) as an "eco-village" (*Shēngtài cūn*), where Ewenki people "live a harmonious relationship with nature." I have conducted interviews with numerous visiting tourists, almost all of them Han-Chinese, who clearly view Ewenki and the region in highly romanticized terms. They commonly use phrases such as "the last primitive tribe" (*Zuìhòu de yuánshǐ bùluò*), the "last hunters," or a "frozen memory of a primitive past." Of course, this is itself interesting as it reflects Han-Chinese, largely urban and middle-class ideas of nostalgia and loss, viewing so-called ethnic lands as unpolluted, natural, and pristine. In this context, Ewenki are expected to perform ethnic culture, sing songs, dance, and demonstrate reindeer skills and crafts. And there is a strong emphasis, particularly in the realm of touristic photography, on happy, smiling femininized faces—and strong, hyper-physical, masculinized bodies.

But it is important to recognize that this is not simply a one-way relation: Ewenki are fully aware of the staged quality of these cultural representations and encounters. And the majority are adept at leveraging the new tourism industry to garner economic benefits. For example, many households earn additional income by offering "tent-stays" in the forest, as well as personalized tours showing so-called "traditional life." Most families have set up shops from inside their homes—selling handicrafts, animal pelts, forest products, and herbs. And, in general, there has been a significant expansion of the reindeer antler market, which has also moved online, as many households have their own Taobao shops (Chinese version of Amazon).

In addition to this, Aoluguya hosted the World Reindeer Herders Congress in 2013, during which China was admitted to membership of the Association of World Reindeer Herders (WRH).

This is an international organization that represents and connects reindeer herding people across the circumpolar North. Founded in 1997 in Nadym, Russia, during the first World Reindeer Herders' Congress, it fosters collaboration, knowledge exchange, and advocacy for the rights, cultures, and traditional livelihoods of reindeer-herding communities. Since the conference, the Chinese state has clearly invested in protecting reindeer and providing support to the community. Besides refurbishing Aoluguya, there have been numerous exchanges between Ewenki and reindeer-herding communities in Siberia and Sápmi—mostly in the form of official visits and research trips from government cadres and academics. I have joined some of these trips and carried out interviews with several members of the Ewenki government, who explicitly describe the Sámi as the best example of reindeer herding in a so-called modern context. Membership to the Association of World Reindeer Herders also brings certain benefits, such as access to the latest veterinary technology and knowledge around breeding. As one might imagine, one of the major problems in the Chinese context is that reindeer have a limited genetic pool because they are cut off from the wild herds to the north. There are thus regular discussions via the WRH to import reindeer from elsewhere in order to increase their numbers (see below).

Running parallel to these official visits and programs are new opportunities for ordinary Ewenki to petition the local government, specifically for cultural and linguistic revitalization. One notable example is language-learning in the local school. Very few young people speak the Ewenki language—and it is officially classified as severely threatened according to the UN. Since 2020, the government has sponsored a teacher to come to the school several times per week to teach language. Interestingly, the teacher is herself a Daur Mongol—who learnt Ewenki growing up in the area—and which aptly captures the problems associated with linguistic loss in the region.

The most ambitious initiative has been a school exchange program with Ewenki communities in Russia. Despite logistical challenges—including the need for special permissions, international travel, and a 24-hour train journey—teachers and students from Aoluguya spent two weeks in an Evenki village in 2018, where reindeer husbandry remains a central way of life. During their stay, students gave presentations, exchanged photographs, and shared glimpses into each other's daily experiences. The visit featured various cultural activities, such as reindeer camp visits, music and dance performances, and the launch of a project to develop a written script for the Ewenki language in China. Teachers also collaborated with their Siberian counterparts to engage in "cultural education," gaining hands-on experience in activities like mushroom foraging, species identification, and forest camping. The exchange was highly successful, leading the local government to expand cultural education in Aoluguya by integrating elders and heritage transmitters into the school curriculum.

In the northeast, many Ewenki have embraced the role of "ethnic entrepreneurs," strategically leveraging their ethnic identity to navigate the market economy. These entrepreneurs operate businesses, such as tourist shops selling cultural handicrafts, or run cultural performance troupes that specialize in traditional music and dance. These troupes perform at various tourist events year-round, with a significant number of young people involved. For the first time, young Ewenki are beginning to see the ethnic performance industry not only as a reliable source of income but also as a means of cultural revitalization. A prime example is the "Aoluguya Musical," a highly successful production written by a prominent reindeer herder. Featuring a dozen Ewenki performers, the musical toured across China for several months, showcasing both the economic potential and cultural value of such initiatives.

These examples prompt intriguing questions about the evolving experience of ethnicity in contemporary China, particularly considering what Comaroff and Comaroff (2009) term "Ethnicity Inc."—a global shift from selling labor to selling culture. Crucially, they challenge the assumption that the commodification of culture necessarily reduces it to superficial forms, contesting earlier anthropological views about the incompatibility of culture and modernity. The experiences of Ewenki appear to support this perspective, as the commercialization of culture has enriched their sense of identity while providing new economic opportunities and platforms for cultural expression.

It is important to remember that, in China, all tourism infrastructure is to some extent state-managed—allowing the state to control the narrative of minority culture, history, and representation. However, what is striking about the new Aoluguya is that Ewenki have retained a significant degree of control over this narrative. This is evident in the new museum, where a prominent museum curator—about whom I have published elsewhere (Fraser 2020)—has written a new ethno-history. His narrative explicitly acknowledges the community's Siberian roots, traces the movement of individual clans over time, and even highlights their Russian affiliations through personal names and religious beliefs.

This shift is also reflected in the museum's cultural content, which includes objects not only from Ewenki in China but also from across the Russian border—emphasizing the enduring cultural connections and flows between China and the Arctic. Similar themes appear throughout the new settlement, where placards and information panels underscore the deep cross-cultural ties between Ewenki and other Indigenous and minority groups across the Circumpolar North.

This transformation in Aoluguya marks a significant departure from past narratives that sought to "nativize" Ewenki as inherently Chinese. Today, their cultural difference and subarctic heritage are not only acknowledged but actively foregrounded. Rather than being assimilated into a singular national identity, Ewenki are now positioned as central to the construction of "Chinese Siberia"—a vision that symbolically blurs the boundary between China and the wider Arctic. In this new framework, their distinctiveness is no longer suppressed but instead heightened and even celebrated. In short, Ewenki help do the symbolic work of constructing the imaginary of China as an Arctic state.

Museums, Photographs, and Heritage

Another apt example of this is in the role of exhibitions. Between 2017 and 2019, I worked on a project involving the Ethel Lindgren Collection, a photographic archive of Ewenki and Orochen communities held at the University of Cambridge. The collection, captured between 1919 and 1921 by anthropologist Ethel Lindgren and her Norwegian husband, photographer Oscar Mamen, offers a rare visual record of northeast China's nomadic Ewenki and Orochen as they hunted and herded reindeer in the taiga. Beyond its historical value, the archive serves as a deeply personal record for present-day Ewenki and Orochen, many of whom see these images as the only visual documentation of their ancestors. The photographs also provide a lens for reflecting on cultural change over the past century (Fraser 2021).

In 2017, a collaboration between the University of Aberdeen, the University of Cambridge, and Inner Mongolia Normal University digitized the archive's several thousand photographs. Digital copies were distributed via hard drives to Ewenki and Orochen minority governments and regional museums. Later that year, I expanded the visual repatriation project in several ways. During ethnographic fieldwork, I used photographs to reconstruct family histories, trace clan migrations, and explore perceptions of social, ecological, and political change, including reactions to policies like the hunting ban.

Another key aspect of my research was following the "afterlives" of the photographs—how they were used (or not) across different networks. I shared the collection with about 70 percent of Ewenki and Orochen settlements, engaging a wide range of stakeholders, including government officials, academics, elders, youth, schools, craft producers, heritage transmitters, and tourism developers. This process highlighted evolving ideas and funding strategies surrounding cultural heritage.

Since then, several things have occurred with the photographs, including museum exhibitions, revitalization of craft work, and even a settlement redevelopment (see Fraser 2021). But most notable for our discussion was an exhibition held at the Aoluguya Museum on the Polar Silk Road. The 2019 exhibition featured photographs, maps, and academic writings on the long history of subarctic cultures in northeast China—set alongside descriptions of China's exploration activities and scientific research in the Arctic. All of this was surrounded by enlarged images of China's *Xue Long* ('Snow Dragon') icebreaker ship, which is often portrayed in news and popular media, and is widely considered a source of national pride. What was most interesting is that the exhibition made no distinction between the political, scientific, infrastructural, and cultural dimensions of the Polar Silk Road; they all intersect and influence one another to help construct the imaginary of China as an Arctic state.

And I argue that we should not underestimate the power of these museological and touristic exchanges in shaping perceptions, narratives, and values. Indeed, a significant number of people from across China visit Aoluguya every year—it is very well known in the popular imagination, as well as on social media. Some Ewenki youngsters have TikTok accounts, showing videos of reindeer herding life in the taiga, and which attract millions of views. And all of this has very real consequences for building the cultural imaginary. I have met countless Chinese tourists in Ewenki areas who have been to Norway, Finland, Russia, and Greenland—or who plan to do so. And, as we know, there has already been a sevenfold increase in Chinese tourism to Arctic states over the past decade.

This is something which will prove particularly relevant in the context of the Polar Silk Road—which is set to initiate increasing tourism but also cultural exchanges between minority groups and Indigenous people (as well as their representative governments, leaders, and intellectuals) along its route. For example, before the outbreak of the Russian war with Ukraine, there were already plans to host a joint Sino-Russian exhibition combining the Lindgren Collection with photographs from the well-known Russian ethnographer Sergei Shirokogoroff—who also visited Ewenki areas during the early 20th century. The exhibition was planned for several sites across the Russian North, including locations along the Polar Silk Road. And interestingly, one of the funders of this exhibition was the Polar Research Institute of China, which is the country's most powerful Arctic research center—focusing on sea ice, glacial monitoring, and atmosphere. So here we see a clear and strategic use of heritage in building relationships between China and neighboring Arctic states, as well as a coupling of scientific research, infrastructure development, and representations of culture.

This is an apt example of what Tim Winter (2016) calls the "cultural diplomacy side of the Silk Road." Indeed, as he points out in the context of the overland parts of the Belt and Road in Central Asia, there is a clear overlap between the sites being developed for Chinese infrastructure and applications for cultural heritage status. Thus, we are now seeing a surge in Silk Road nominations at UNESCO's annual world heritage committee meetings—precisely as Chinese investment is moving through that region. As Winter writes, given that there are dozens of Silk Road corridors linking more than 500 sites, the Belt and Road is likely to emerge as "the most ambitious and expansive international cooperation program for heritage preservation ever undertaken" (2016: 1).

I am particularly interested in how this might manifest in the context of the Polar Silk Road. Indeed, it is not hard to imagine that—just as with the historical photographs mentioned above—the Indigenous and ethnic minority communities in the Arctic will become entangled in similar processes of cultural heritage diplomacy. Who is to say that photographic archives of reindeer herding and hunting communities such as Ewenki, traditional objects, and material culture from Siberian Indigenous peoples, but also traditional Sámi craftwork such as duodji, will not also become tools in the future politics of cultural heritage applications?

And of course, this raises many important questions: such as what kinds of cultural heritage might or might not be included? Who will speak for—and on behalf of—local communities, heritage transmitters, and other relevant stakeholders? How will traditional objects, photographs, and material culture be represented and managed? And what role will museums, researchers, and possibly activists play? As noted by several scholars (Hafstein 2012; Beardslee 2015; Coombe 2015), classifications of cultural heritage are not a neutral or apolitical celebration of ethnic traditions. Instead, they are a selective process that reinforces specific ideas and values, while shaping the ways in which culture can be represented (see also Fraser 2020).

This is especially evident in the Chinese context (Svensson and Maags 2018; Blumenfield and Silverman 2013), where cultural heritage often serves as a subtle method of "rendering individuals governable." It plays a role in managing and controlling tradition, defining cultural practices, and guiding people's memories, senses of place, and identities in specific directions (Beardslee 2015). And, as already alluded to in the photographic example above, cultural heritage also reproduces existing political hierarchies. Afterall, heritage often remains in the hands of powerful local actors, what Beardslee (2015) calls "heritage middle-men," who not only operate with relative flexibility in harnessing channels of funding to promote cultural revitalization, but which also strengthens their own positioning within the local politics of representation (Fraser 2020, 2022). Given the immense transformations expected from China's pivot to the Arctic, we should not underestimate the role of these issues—and the wider potential played by heritage and culture across the region. As Xi Jinping has himself stated: the Belt and Road is not just about trade—it is about culture. In his words, it is "an initiative to promote inter-civilization and cultural exchanges, to build bridges of friendship, drive human development and safeguard peace in the world"(The Third Belt and Road Forum for International Cooperation 2023). Now these are predictably bold statements, but they should show that China is taking all of this very seriously; that culture and heritage are at the heart of the Belt and Road as forms of soft power—and this will certainly also be so in the context of the Arctic as well.

Conclusion

This raises one final dimension that I would like to end with, namely, the side effects or unintended consequences of the Polar Silk Road—especially in terms of building relationships and networks between stakeholders along its route. As many of my Ewenki friends have stated, for them the Polar Silk Road is not about infrastructure and resources but an opportunity to interact with—and to learn from—other northern minorities and Indigenous people who have had comparable experiences of colonization, assimilation, and cultural-linguistic loss. Everyone knows that in Aoluguya today, there remain deep social problems left behind from the legacies of sedentarization, relocation, and the hunting ban. Very few young people speak Ewenki and Orochen languages. There is persistent alcohol dependency amongst the older generations and, despite recent developments in the context of heritage and ethnic tourism, it remains to be seen exactly how these will help with genuine cultural revitalization and meaning-making—particularly amongst the next generation.

For the first time ever, Ewenki have access to significant levels of funding and state support, but as they themselves state it sometimes feels as though it is too late; what they require now is direct engagement with—and experience from—other communities and places, and the Polar Silk Road is seen as the best resource to achieving that. This is something that has again connected me to the story. Just a few months ago, the Ewenki government approached me (and my colleague in Finland) to assist in purchasing several hundred reindeer from neighboring Siberia—their goal being to increase the reindeer population in Aoluguya. This is logistically highly complex—and involves transporting live reindeer thousands of kilometers by train across the international border. Since then, discussions have progressed, and next summer, the state plans to transport 250 reindeer from Yakutia. While this initiative has all the hallmarks of a statist Maoist-era campaign, it signals the willingness of the Ewenki government to harness the opportunities brought about by the cultural imaginary of China as a near-Arctic state. As several of my friends have joked, it signals the building of *their* Polar Silk Road. So, what I am interested in is what kinds of linkages and networks might emerge—how people might build a parallel Polar Silk Road of ideas—culture, heritage, and collaboration—and, particularly in the context of the climate emergency—quite possibly activism and cultural revitalization as well.

Richard Fraser is an Associate Professor within the Anthropology Section of the Department of Social Sciences at UiT—the Arctic University of Norway. He has carried out long-term ethnographic fieldwork with Ewenki, Orochen, and Mongol communities in northeast China since 2007, as well as with Darhad and Tsaatan communities in Mongolia since 2010. More recently, he has started new fieldwork in China's Southwest, in Guizhou and Sichuan provinces. His main areas of research are environmental, multi-species and more-than-human relations, multimodal anthropology and critical heritage studies, and the changing role of museums, digital sharing, and archives for minority-state dynamics.

Note

1. For a discussion of a comparable museum in the Orochen Autonomous Banner see Fraser (2022).

References

Althusser, Louis. 2014. *On the Reproduction of Capitalism: Ideology and ideological State Apparatuses*. London and New York: Verso Books.

Anderson, David. 2000. *Identity and Ecology in Arctic Siberia: The Number One Reindeer Brigade*. Oxford: Oxford University Press.

Beach, Hugh. 2003. "Milk and Antlers: Chinese Dual-Ownership System Remains a Hopeful Model Despite Forced Relocation from Olguya (Inner Mongolia)." *Cultural Survival Quarterly* 27(1): 33–35.

Beardslee, Thomas. 2015. "Whom Does Heritage Empower, and Whom Does It Silence? Intangible Cultural Heritage at the Jemaa El Fnaa, Marrakech." *International Journal of Heritage Studies* 22(2): 89–101.

Blumenfield, Tami, and Helaine Silverman, eds. 2013. *Cultural Heritage Politics in China*. New York: Springer.

Bulag, Uradyn. 2010. *Collaborative Nationalism: The Politics of Friendship on China's Mongolian Frontier*. Lanham: Rowman & Littlefield Publishers.

Castoriadis, Cornelius. 1987. *The Imaginary Institution of Society*. Translated by Kathleen Blamey. Cambridge: MIT Press.

Comaroff, John, and Jean Comaroff. 2009. *Ethnicity Inc*. Chicago: University of Chicago Press.

Coombe, Rosemary. 2015. "Managing Cultural Heritage as Neoliberal Governmentality." In *Heritage Regimes and the State*, ed. Regina F. Bendix, Aditya Eggert, and Arnika Peselmann, 375–87. Göttingen: Göttingen University Press.

Dumont, Aurore. 2017. "Declining Ewenki 'Identities': Playing with Loyalty in Modern and Contemporary China." *History and Anthropology* 28(4): 515–30.

Elliott, Mark. 2001. *The Manchu Way. The Eight Banners and Ethnic Identity in Late Imperial China*. Stanford: Stanford University Press.

Fraser, Richard. 2010. "Forced Relocation amongst the Reindeer-Ewenki of Inner Mongolia." *Inner Asia* 12(2): 317–46.

———. 2015. "Tents, Taiga, and Tourist Parks." In *Reclaiming the Forest: The Ewenki Reindeer Herders of Aoluguya*, ed. Åshild, Kolås, and Yuanyuan Xie, 130. New York and London: Berghahn Books.

———. 2020. "Cultural Heritage, Ethnic Tourism, and Minority-State Relations amongst the Orochen in North-East China." *International Journal of Heritage Studies* 26(2): 178–200.
———. 2021. "Contesting the Chinese Taiga: Spirits, Reindeer, and Environmental Conservation in Northeast China." In *Cosmopolitical Ecologies across Asia: Places and Practices of Power in Changing Environments*, ed. Riamsara Kuyakanon, Hildegard Diemberger, and David Sneath, 184–204. London: Routledge.
———. 2022. "A Museum in the Taiga: Heritage, Ethnic Tourism and Museum-Building amongst the Orochen in Northeast China." *Journal of Heritage Tourism* 17(5): 493–515.
———. Forthcoming. 2026. "Antlers, Organs, and Spirits: Ontological Frictions and the Biopolitics of Reindeer Parts in Northeast China". *Inner Asia*.
Hafstein, Valdimar. 2012. "Cultural Heritage." In *A Companion to Folklore*, ed. Regina Bendix and Galit Hasan-Rokem, 500–19. West Sussex: Chichester Wiley-Blackwell.
Kim, Loretta. 2009. "Marginal Constituencies: Qing Borderland Policies and Vernacular Histories of Five Tribes on the Sino-Russian Frontier." Ph.D. dissertation. Cambridge: Harvard University.
Kolås, Åshild. (2011) "Reclaiming the Forest: Ewenki Reindeer Herding as Exception." *Human Organization* 70(4): 397–404.
Lindgren, Ethel John. 1933. "Northwest Manchuria and the Reindeer-Tungus." *The Geographical Journal* 75(6): 518–34.
Lundberg, Maria, and Yong Zhou. 2009. "Hunting-Prohibition in the Hunters Autonomous Area: Legal Rights of Oroqen People and the Implementation of Regional National Autonomy Law." *International Journal on Minority and Group Rights* 16(3): 349–97.
Svensson, Marina, and Christina Maags. 2018. *Chinese Cultural Heritage in the Making: Experiences, Negotiations and Contestations*. Amsterdam: Amsterdam University Press.
Ma, Rong, 2007. "A New Perspective in Guiding Ethnic Relations in the Twenty-First Century: 'De-politicization' of Ethnicity in China." *Asian Ethnicity* 8(3): 199–217.
Mullaney, Thomas. 2011. *Coming to Terms with the Nation. Ethnic Classification in Modern China*. Berkeley: University of California Press.
Nentwig, Ingo. 2003. "Reminiscences about the Reindeer Herders of China." *Cultural Survival Quarterly* 27(1): 36–38.
Oakes, Timothy. 2005. *Tourism and Modernity in China*. London and New York: Routledge.
———. 2013. "Heritage as Improvement: Cultural Display and Contested Governance in Rural China." *Modern China* 39 (4): 380–407.
Stammler, Florian, and Hugh Beach. 2006. "Human–Animal Relations in Pastoralism." *Nomadic Peoples* 10(2): 6–25.
The Arctic Institute – Center for Circumpolar Security Studies. 2022. *China*. https://www.thearcticinstitute.org/country-backgrounders/china/.
The Third Belt and Road Forum for International Cooperation. 2023. *A Global Community of Shared Future: China's Proposals and Actions*. http://www.beltandroadforum.org/english/n101/2023/1010/c127-916.html.

Winter, Tim. 2016. "One Belt, One Road, One Heritage: Cultural Diplomacy and the Silk Road." *The Diplomat* 29: 1–5.

———. 2021. "Geocultural Power: China's Belt and Road Initiative." *Geopolitics* 26(5):1376–99.

Xinhua. 2003. "Last Hunting Tribe Gives Up Virgin Forest." August 8. Xinhua News Agency.

Yeh, Emily. 2005. "Green Governmentality and Pastoralism in Western China: 'Converting Pastures to Grasslands.'" *Nomadic Peoples* 9: 9–22.

Chapter 2

Vast Spaces and Inconsistent Anticipations in the Northeast Siberian Arctic

Florian Stammler, Aytalina Ivanova, and Piers Vitebsky

Some visions of the future are easier to anticipate than others. Many regions of the world have now undergone increasing population for centuries. A defining feature of global modernity is the assumption of an inevitable increase in size, scale, and activity— "growth," with its partner "development" (Hobart 1993). Modernist assumptions of such unending development create forms of anticipation which are expansionist in principle.

But this kind of scenario is not historically inevitable. Every site excavated by archaeologists—even capitals of great empires—was once abandoned as no longer viable or relevant, at least in that form at that time. In some places this process is happening in front of our eyes today.

Shrinkage in a managed, positive form is emerging as a general development policy in many regions. Thus, Richardson and Nam (2014) include case studies from India, China, Latin America, the US, and the EU. Here, however, we shall look at sites where degrowth has happened unintentionally, with authorities and people left to grapple with the consequences alone, even to the brink of abandonment. Recent Arctic research has emphasized the agency of people in such places, when shrinking becomes part of a branding exercise

(Gansauer, Lilius, and Adams 2025), and inhabitants self-organize as the state retreats from their places (Grossmann et al. 2024). Bernt et al. (2014) have emphasized that part of the problem is that growth remains the continuing ideal. Grossmann et al. (2024) emphasize the risk that even scholarship on shrinking cities perpetuates such neoliberal development agendas. How do local people organize their lives and make plans in such a context? Are they likely to adopt Indigenous cosmologies that are better suited to post-development agendas? In the Arctic and elsewhere, Indigenous people may feel more at home in a cyclical cosmology, where linear development, let alone growth, is not even a relevant concept (Stammler and Ivanova 2016b: 61).

Many settlements in the Russian Arctic, from the largest industrial city to the tiniest mining camp or Indigenous reindeer farm, were artificially created in the 20th century to fulfil a policy of the Soviet state. When that rationale ceases to be valid, these settlements atrophy. Like other industrialized societies, the Soviet Union presupposed an indefinite expansion of settlement and resource extraction, with the additional twist that space was indefinitely available, so that rather than clean up old wreckage it was often simpler to start again in a new place.

Since around 1990, much of the Russian Arctic has experienced drastic contraction and population exodus, particularly in the northeast. Settlements in this region are characterized by some specifics: structural disadvantages due to remoteness, a harsh climate, specific processes of place-making related to industrial history over less than 100 years (Bolotova and Stammler 2010), and the presence of Indigenous minorities in the surrounding tundra (Stammler and Khlinovskaya-Rockhill 2011). This makes them significantly different from shrinking cities in post-socialist Europe (Haase et al. 2016). Hill and Gaddy's influential but deeply flawed volume (2003) argued that the Russian Arctic is over-populated. This would become the dominant view as planners in far-away Moscow or the World Bank in Washington saw a "surplus" population as a burden on supply lines (Heleniak 2001). Since then, scholarship has significantly contributed to a more fine-grained analysis of social processes not only among the leavers (Heleniak 2009; Nuykina 2011), but also among the stayers in the region (Orttung 2016). Particularly, ethnographic research has revealed the varied and subtle forms of people's attachment to the region (Thompson 2008; Bolotova and Stammler 2010; Stammler and Khlinovskaya Rockhill 2011). These works emphasize how such policies fail to understand local senses of place-making. Moreover, recent scholarship within the Russian North and beyond has emphasized how those who stay in shrinking or crisis-stricken places are not just passive victims of policies but display agency to keep their places viable (Nuykina 2014; Adams and Komu 2021; Adams, Allemann, and Tynkkynen 2022; Bolotova 2022), even turning their shrinking into an advertisable brand (Adams, Bolotova, and Alasalmi 2023; Gansauer, Lilius, and Adams 2025).

In their study of the perception of mining prospects among villagers in northern Sweden, Anshelm and Haikola (2017: 21) document rises and falls in what they call "horizons of expectation." We believe that this idea can be extended, by examining local social relations and processes, to encompass more specific horizons of mood: of frustration or disappointment, cynicism or resignation—and of unofficial local opportunism. Anticipation takes place within a social and historical context of agendas, trajectories, extrapolations, and struggles. There are not only anticipations in the abstract, but actual people who do the anticipating, who feel and who plan.

The Northern Sea Route (NSR) runs along the Arctic coast linking Murmansk and Arkhangelsk in the west with Vladivostok, Japan, and China, and provides substantial savings in distance for cargo ships running between northwestern Europe and the Far East. Within the overarching context of Russian security concerns and Chinese investment interest, we shall look at part of this route along the remote Arctic Ocean coast of the Sakha Republic (Yakutia) in the Siberian northeast, and in particular at some enterprising personalities in two strategically located small settlements. As the population in the empty, darkened apartment blocks of the sea port of Tiksi shrinks from 12,000 in 1989 to 4,000 in 2021, others plan its expansion as a vital support station on the NSR between China and Europe. Naybá, a tiny outpost of some 150 Indigenous reindeer herders 112 km eastward along the same coast, is even more remote yet with the geographic potential to challenge Tiksi as a future international deep-water port. Both settlements are extremely isolated and hard to reach, with no road to the bustling regional capital Yakutsk 1,000 km to the south, a diamond boom city linked to China, and until 2022 to Japan and South Korea, in high-velocity flows of business, education and consumption.

Northeastern Siberia and Chinese Interests

The Sakha Republic (Yakutia) occupies a vast space. It is approximately the size of India but contains a little under 1 million people. As a federal republic it has its own regional government with a full range of ministries, though these are ultimately subordinate to the government of the Russian Federation in Moscow.

Small settlements of Indigenous minorities (Evenki, Eveny, Yukaghir, and Chukchi) are thinly scattered across the huge expanses of the north, occupied mostly in reindeer herding supplemented by subsistence fishing and hunting. Before its shrinking, Tiksi used to be a remote "Russian" outlier of aviators, navigators, and explorers along the Soviet Northern Sea Route. The city of Yakutsk is the cultural and political center of the Sakha ("Yakut"), the dominant local Asian people, whose traditional culture was based on herding horses and cattle. Settlements of "Russians" (a term including other ethnically European citizens of the Soviet Union such as Ukrainians, Estonians, etc.) are concentrated to the south of the Republic in areas of coal, gold, and diamonds, where they work as miners, engineers, and technicians.

In the 1990s, Yakutsk developed rapidly as the region's Sakha population moved from nearby depressed rural areas to the city (Vitebsky 2000) and grew from 159,000 in 1980 to 361,000 in 2023. A mushrooming of government facilities, apartment blocks, shiny glass business centers, and university buildings was financed by revenue from natural resources. This high point of regional autonomy, during which a substantial proportion of diamond revenue returned to Yakutsk rather than staying in Moscow (Tichotsky 2000; Argounova 2004), was restricted after 2000 with the increasing assertion of centralized control (Ivanova and Stammler 2017).

Transportation to the outside world remains uneven. The main route from Yakutsk to elsewhere remains through aviation. Residents can fly for business, holidays or medical treatment to Moscow direct, and with a change to Beijing, Harbin and until 2022 to London, Tokyo, or Seoul. A branch line from the Trans-Siberian railway was extended in 2019 to Nizhnyi Bestyakh, on the opposite bank of the huge River Lena facing Yakutsk itself. The road parallel to the railway (A360 Lena) is now largely paved, which will likely change the character of narratives and local practices (Argounova and Prisyazhnyi 2016). After seemingly eternal delays, many had stopped believing that a bridge crossing the Lena River would eventually be built to connect the only metropolis of this huge territory to the road network beyond. But as of March 2025, with active use of Chinese equipment, construction was in full swing. Foundations have been drilled already on both shores and concrete factories installed for the giant pillars of the 4.6 km long bridge.

This will end the city of Yakutsk being cut off from ground transportation except by driving across the ice in winter, making helicopter and hovercraft the only link at other times to the infrastructure across the river. This bridge confirms the tendency of projects with Chinese investment in the Russian Arctic to remain at the stage of anticipation. While the Chairman of Hua Tsing An Zu Company said ten years ago: "We are highly interested in the project. We could build a bridge in three and a half years. [Yakutia] can become a driving force of the economic zone along the Silk Road"[1]—today there is no evidence of Chinese money in the project. When the bridge connecting Crimea to the Russian mainland became a priority, many considered the Lena bridge even less realistic, although authorities publicly kept promising its completion by 2028.[2] Now it seems to be happening not because of, but in spite of a lack of Chinese investment, even though Chinese representatives keep emphasizing how they are committed to the development of the Arctic and emphasize the importance of Yakutia for their agenda.[3] We shall briefly review the substance and tone of these anticipations of Chinese investment, before focusing on how these are refracted in our two small settlements on the remote coast far to the north.

Few of these anticipations of Chinese investment are fulfilled. Two flagship oil and gas projects, the East Siberia Pacific Ocean oil pipeline and the Sila Sibiri (Power of Siberia) gas pipeline (Sidortsov et al. 2016), actually deliver Siberian oil and gas to China, 23 billion m^3 gas in 2023. However, Russia's hope for Chinese investments in a second pipeline (Power of Siberia 2), compensating for the losses of the European gas market for Yamal gas, has not (yet) materialized.[4] The first Chinese company announced their investment in the Sakha gas business, by buying a 10 percent stake in YaTEK company, but there is no evidence again about the deal progressing beyond the memorandum of intention from early 2022. Nonetheless, YaTEK apparently delivered a test trainload of LPG (liquefied petroleum gas) to China in 2023.[5] Similarly, the tangible consequences of the 50 percent stake of the Chinese businessman Li Duan in the Sila Yakutii (Power of Yakutia) company remains unclear. The plan to export electricity from Yakutia to China and build an 8 gigawatt gas energy power station in Lensk, with a connecting powerline from Lensk to Blagoveschensk, remains at the stage of a memorandum too.[6]

The numerous projects with China, both realized and intended or anticipated, show that the main interest is in the trade in raw materials from Siberia to China and of manufactured goods the other way. Consumer goods and food are imported from China through the Amur region by road, crossing the Amur river at the final point, since June 2022 on a bridge built by Chinese from the southern side and

Russians from the northern side.[7] The big Moscow-operated supermarket chains (Magnit, Pyaterochka) have not reached Yakutsk, and the city remains dominated by the Chinese market in wholesale and retail, even though formally Chinese citizens are now forbidden to have their own businesses.

These are local examples of the general Russian–Chinese cooperation that includes but is not limited to the Arctic, for example as part of the "One Belt One Road" initiative. During Xi Jing Ping's visit to Moscow in March 2023, they agreed on establishing "a joint working organ" on the development of transit on the NSR.[8] This may revive another beautifully named Chinese idea with Russia, the "Polar Silk Road" (*Ledyanoy Shelkovyi Put'*), part of the "One Belt One Road" initiative which was proudly announced by Putin and Xi Jin Ping in June 2017, and which leaders continue to raise anticipation about.[9]

Despite some real investment or delivery of the projects mentioned, most promised Chinese–Russian projects in Yakutia fail to materialize: in 2015 the regionally owned oil company Tuymaadaneft agreed with a Chinese company to build an oil refinery in the South Yakutian district of Aldan.[10] This oil refinery was not built, nor was one of Yakutsk's neighborhoods built and populated by Chinese, nor were Chinese bricks produced in the special economic zone of Kangalassy, nor was Chinese investment in the Northern Sea Route infrastructure realized, nor was the world's biggest gas power station built in Lensk (see above), nor was a railroad built connecting Yakutia to the Chinese town of Mokhe, to name a few more examples that were once confidently announced.

In Yakutsk the Chinese presence evokes an ambivalent attitude. Promises of developments are officially welcomed, and sustain a momentum of anticipation. But many in the population fear Chinese immigration and complain about being flooded with low-quality goods from China sold expensively on the captive market in Yakutsk, and the predatory extraction of Yakutia's natural resources. One Sakha analyst argues that by signing memoranda of understanding and preliminary agreements in Yakutia, Chinese private companies attract investment funding from the Chinese state, which they subsequently re-route into projects within China that promise better revenue. In such a view, signing empty memoranda appears to be intentional on the Chinese side.[11]

Ultimately, significant investment in infrastructure still tends to come from the Russian government, such as the Lena River bridge and the new ship-wharf in the Yakutsk suburb of Zhatay. Since the Lena River flows downstream to Tiksi, leading to the Laptev Sea and the NSR, this wharf is of crucial importance for all of East Siberia. Therefore Moscow included it in their state program for the "development of shipbuilding and technology for the development of shelf deposits," with a budget of 5.75 billion rubles, of which 4 billion was allocated from the 2019 Federal budget. As of March 2025, construction is almost completed, and the wharf should start building the first vessel in 2025. This seems to be achieved also thanks to import of Chinese high-tech to replace western ship-welding technology.[12]

The Arctic Ocean Port of Tiksi

Far from the buzz of Yakutsk, the settlements on the Arctic Ocean coast seem like a different world.

This is a federally controlled "border zone," where any visitor, even a resident of other nearby areas, needs a special permit from the Federal border guards to enter. Though many residents of Yakutsk fly to Moscow or foreign countries, very few indeed have visited this region or have much idea of life here.

The town of Tiksi, on the huge delta of the Lena River which flows northward from Yakutsk, was established in 1933 as a key stopping point on the Northern Sea Route. Until very recently ships moved only in summer under the escort of Russian nuclear-powered icebreakers. Since 1991 this service has also been available to foreign cargo ships and the Russian government hopes to earn considerable revenue from this, though currently the main foreign interest is from China, leading to vast horizons of expectation among Russian scientists as they predict 180 billion ruble investments in the NSR in the near future (Leonov and Zaostrovskikh 2021). Since with recent melting, nuclear icebreaker services may become less necessary in summer, the plan is to offer their services for keeping the NSR open round the year, with new investments in even more powerful icebreakers.

In the early 20th century, the NSR was a key driver of the socio-economic development (*osvoenie*, "mastery") of the Russian Arctic. In the early Soviet period, the NSR had its own dedicated ministry (Glavsevmorput, short for "Great Northern Sea Route"), responsible for planning infrastructure. The route functioned for intra-Soviet transportation, and by its Soviet-time height in 1987 it was served by 17 icebreaker vessels, transporting 6–8 million tons of cargo per year, significantly less than the 30 million tons of 2021[13]. Supplies of food, machinery, heavy plant, spare parts, and fuel oil are still brought to ports at the mouths of Russia's northward-flowing rivers and trans-shipped onto river craft for the journey upstream to towns and cities (such as Yakutsk) in the south.

Though the NSR is administered overall from the Russian western Arctic port town of Murmansk, Tiksi continues to host the port administration, the navigation, search and rescue facilities for its region, as well as the administration of the Bulunski district of the Sakha Republic. Moreover, an important division of the Russian army in the Arctic is based in Tiksi, in a suburb referred to as Tiksi 3, some 7 km from the main town.

Until the mid-1990s the NSR was a major supply route for the city of Yakutsk, with cargo trans-shipped at Tiksi on to the river fleet for the journey upstream. Many important goods arrived in Tiksi first, leading to a local pride about their privileged position in the Soviet distribution system, through their direct maritime connections to Murmansk and Arkhangelsk and air to Moscow: "In the early 1990s we were the first ones to get cargo ships with Japanese cars. We were living better than Yakutsk back then."

Even though the NSR opened for international navigation in 1991, the infrastructure along the coast has been stagnating since the 1990s, much to the regret of local residents. The dependence of Yakutsk on Tiksi has been greatly reduced, through the overall decline of the shipping enterprise, plus a backlog during the 1990s of bills unpaid to the shippers by the Sakha authorities. This situation is generally masked by the buoyant supply situation of consumer goods and machinery in Yakutsk city, but these are increasingly brought in through a combination of air cargo from Moscow to the enlarged airport in Yakutsk, or rail trans-shipment on the trans-Siberian railway and the Baikal-Amur rail to Nerungryi and from there since 2019 northwards by road and rail to Nizhnyi Bestyakh on the shore of the Lena river opposite Yakutsk.

After a high point in 1989 with a population of over 12,000, Tiksi experienced the exact opposite of Yakutsk—heavy population decline—and has now approximately 4,000 inhabitants. Most of those who left were of Slavic origin, as in the other industrial cities in the Russian North, while Sakha and Indigenous inhabitants of small settlements moved to some of the empty apartment blocks left behind by those whom they generalizingly call "Russians." In 2012 the army left, abandoning any support for the airport, and Tiksi 3 became a ghost town. For our first fieldwork there in summer 2012, it was very hard to get to Tiksi even from Yakutsk, with only helicopters and small airplanes able to land on half-destroyed runways with an interim permit. That same winter Tiksi almost froze to death when the local heating plant collapsed in the middle of the Arctic winter. Almost ten years later, our host in Tiksi remembered: “We were grouping together with neighbors in somebody’s apartment, and tried to keep the plumbing and radiators from freezing by wrapping them in blankets and everything we could find. It was a fight against time. We needed to keep our pipes intact while they were restoring the warm water supply in the central powerplant. And we managed!” The pressure was on the state not to let its remote citizens freeze to death in the Arctic winter, as the powerplant incident had put Tiksi at the top of the headlines all over Russia.

While short-term outsiders generally see ruination, on the inside there is a great deal of positive activity. Local people completely refurbished and enlarged the culture and sports center (Kul’turno Sportivnyi Kompleks, KSK). Rather than a conventional “House of Culture” (Donahoe and Habeck 2011), the new director, Vasily Burtsev, united state-of-the-art sports and exercise equipment with other sociocultural activities, offering the institution as a meeting place for young people and social innovation. One such initiative was the “Iron Will Team,” who dedicated their energy to promoting a “healthy way of life” (*zdorovy obraz zhizni, ZOZh*) to young people, and combating alcoholism and violence. Burtsev and his team at KSK connected with cinema enthusiasts and youth movements all over Russia, obtained grants, and established an IMAX cinema in Tiksi even before one reached Yakutsk. They also cleared a huge amount of old industrial junk, from both land and sea. As a legacy of the Soviet tendency to move elsewhere rather than to clear up, the bay is filled with rotting timber and wrecks of old Soviet vessels. In 2022, Vasily Burtsev contributed to the national volunteer campaign “A Clean Arctic,” where volunteers collected 800 tons of trash from Tiksi alone.[14] We saw Tiksi airport being revitalized in 2021, promising to restore Soviet-era direct connections to Moscow, from a time when those connections were valued more than those to Yakutsk (Stammler, Ivanova, and Sidorova 2017). However, these hopes have yet to materialize, though social media are full of such anticipations[15].

The most tangible signs of development are the buildings: during our first fieldwork in 2012 (Stammler et al. 2017) the Orthodox church was under construction, and the culture and sports center KSK was brand new. In 2021 construction was also going on by the Russian meteorological service (Rosgidromet), which restored one of the big Soviet houses in central Tiksi and turned it into their main Eastern Arctic headquarters (figure 2.1, completed as of 2025).

Figure 2.1. *Renovation of a half-abandoned Soviet apartment block to host the Russian Meteorological Service's Eastern Arctic Headquarters, 2021. © Aytalina Ivanova.*

Local people expect that more office space for other companies will also become available there.

With the increasing militarization and securitization of the Russian Arctic (Ivanova and Fondahl 2020), the Russian border guards increased their presence and started building a new house for their staff in the central settlement. Among the half-ruined Soviet buildings, the construction site in 2019 already looked impressive (though the building was still not livable in late 2021, while the cost had risen to 1 billion rubles). After a six-year period as ghost town, the military settlement of Tiksi 3 became an important Arctic base for the Russian army again in 2019, when the new anti-air attack base of the 45th army was established there, with plans to further enlarge the operations. The underlying rationale is that the resources of the Arctic, including shelf deposits, must be protected from external threats, for example the US, as they are geographically not far away.[16] During our fieldwork in 2019 and 2021, warships were navigating directly between Murmansk and Tiksi along the NSR and in Tiksi we noticed the large number of uniformed soldiers on the roads. Vasily Burtsev confirmed that soldiers now make up half of the clients of his sports facilities.

At the same time, in the civilian town of Tiksi there was discontent at how the army invested only for their own benefit and not the rest of the town. Rumors went around that the army would give their half-abandoned town of Tiksi 3 a state-of-the-art internet connection from an optical fiber cable that was going to be put on the sea bed along the NSR from Arkhangelsk. In town people were apprehensive that the army would not invest in the seven additional kilometers of the cable to connect the main settlement of Tiksi. However, in March 2022, the entire project was put on hold, because of the withdrawal of Finnish and Japanese investments in part of the NSR internet cable. The hope was then placed on investments from the Russian state, but the project has not advanced since then.[17]

In August 2022 the Russian government ratified their NSR development plan until 2035, which includes the modernization of Tiksi as a deep-water sea port, as well as the construction of a search and rescue station of the Ministry of Emergency Situations (the other such stations being in Sabetta, Dikson, and Pevek). Though there is no concrete mention of budget and timings for the port construction, the search and rescue facilities already have budget allocations and timings, and are due to be completed by 2029. Part of the plan is also the re-creation of the early Soviet Federal NSR administration, even reviving the same name—Glavsevmorput—under the leadership of the state nuclear company Rosatom, which has a regularly updated website in Russian and English under this historic name.[18]

The NSR itself, and the role of deep-water ports, depends on global climate shifts. Here, too, local people have their own anticipations. As everywhere in the Russian Arctic, everyone is very well-read in popular (and sometimes technical) science. Some activists also engage in writing themselves, such as the Tiksi biologist and former Director of the Lena Biosphere Reserve Aleksandr Gukov. He is also a passionate local historian who has published his own massive biography of his town (Gukov 2013). He runs a private museum, which has a room dedicated to global warming, where he has collected what he refers to as all the available evidence to prove the cyclical nature of warming, claiming that none of it is man-made. He anticipates that within the next thirty years a new little ice age will come over us, hence we should extract all possible oil and gas resources from the Laptev Sea now, before this becomes more difficult. He interprets the securing of oil and gas licenses by Rosneft as a hopeful sign that this development is about to happen, with Tiksi becoming the main hub.

Thus, the departure of the army, the abandonment of the airport, and the failure of the heating plant marked a trough of decay which people like Aleksandr think is now being reversed. In people's memory of that time, among those who did not leave there is still a sense of pride that they endured these hardships and survived. And indeed, the airport has been re-opened, an additional runway is being built, the army has come back, and several Russian investment projects have brought significant local construction in the civilian part of Tiksi too.

While our conversations in Tiksi show that the most tangible results of development are linked to domestic Russian projects, it is still the international implications of the NSR and related prospects that evoke the most elaborate anticipations or horizons of expectation. Yet there is a problem: China's main interest on this coast seems to be in the shipment of liquefied natural gas (LNG) from the newly built port of Sabetta on Yamal Peninsula to the port of Jiangsu Zhudong in China. With climate change, this needs less and less icebreaker support, with the first vessel already sailing independently in 2018. Therefore, Russia now promotes the NSR as a year-round navigation route. As a result, the fleet of nuclear icebreakers stationed in Murmansk would no longer be essential, nor would supporting infrastructure along the coast of the NSR, including Tiksi.

In addition, Tiksi suffers from a downside aspect of the very same river which led to its creation in the first place as a trans-shipment point between the Arctic Ocean and the River Lena. Local people remember interesting encounters with crews of foreign ships in the past, bartering skins for knives, but recently we saw big international vessels passing by far out along the Northern Sea Route. Though domestic and military vessels call in, Tiksi is too shallow for proper high ocean ships to land because of the huge deposits of sediment that the Lena brings down to the Laptev Sea. Every time the water is dredged and deepened, sediment is washed in again. If large ships do stop at Tiksi, they do so far out from the shore, and the cargo needs to be re-loaded on to smaller vessels. If only big vessels could land, lamented Vasily Burtsev, Tiksi would be a world transport hub and "we would be living like in Dubai."

The Tiny Reindeer Herding Village of Naybá

Any expectations for the port of Tiksi come up against the shallow water, the industrial junk, and the regular need to clean up the river sediment. However, Tiksi's horizon of frustration opens up a further horizon of anticipation and opportunity in Naybá, a tiny settlement mostly of Indigenous Eveny. Here, some people have high hopes of hosting the NSR deep-water port instead of Tiksi. This natural bay along the coast is far from the silt of the Lena River, with no trash from ruined Soviet infrastructure to be tidied up before anything can be built. Naybá, like Tiksi, is still accessible for dual-purpose river ships of the class *Reka-More* ("River-Sea"), but in Naybá high ocean ships could also dock at the port right by the shore (figure 2.2).

Officially, there are 473 residents registered in Naybá, though the actual population is probably around 150. The intimate arena of Naybá provides a striking example of the interaction of anticipation and individual entrepreneurial personalities.

Like other northern villages, Naybá inherited a dual structure from the Soviet Union, where authority was divided between the elected head of the village's civil administration, and the head of the "state farm" (*sovkhoz*), now often re-branded as a cooperative "community" (*obschina*)—though Naybá has become a Municipal Unitary Enterprise (MUP) since Vladimir Alexeyev believes the domain of agriculture attracts better funding than that of indigeneity. The MUP amounts to the village's total productive economy, based mainly on reindeer herding. The relationship between two such leaders can be

Figure 2.2. *The tiny village of Naybá on the shore of the Laptev Sea, 2019. © Florian Stammler.*

either harmonious or conflictual. Vitebsky (2005) describes another Eveny village which was paralyzed in the 1980s–90s by a conflict between a dynamic, progressive village head and a reactionary sovkhoz director who systematically thwarted his plans for community improvement. Naybá is fortunate in the agency-enhancing synergy of its two leaders. The administrative head at the time of our fieldwork in 2021, Vasily Balanov, is an eloquent communicator, and passionate about giving his village a vibrant local economy with good global connections. His counterpart, the MUP head Vladimir Alexeyev, is much less of an orator but a gifted master of all kinds of skills and a good businessman who tries hard to create favorable working conditions for the reindeer herders in his community. He invests part of his own business income to accomplish this, alongside a shrewd harnessing of resources from the state, much as described earlier for other communities in Siberia and the North by Humphrey (1998) and Konstantinov (2000, 2002).

The range of possible anticipations and actualizations is highly circumscribed by official agencies and their policies and prejudices. Community leaders like Vasily and Vladimir must work against the inertia of a system in which goals, policies, and resources are largely located far away and at higher levels of government (cf. Khlinovskaya Rockhill, Sidorova, and Vitebsky 2022, where the normative structure of the nomadic family is imagined by city officials ignorant of local conditions).

Thus, the village's name *Naybá* in the Eveny language means Arctic char (*Salvelinus alpinus*)—an obvious indicator of a key local resource. Yet offshore fishing was completely forbidden. Before 2012, Vasily says, nobody checked here anyway, so there was no need to have a license to fish. But then the *Rybinspektsiya* (fish inspectorate) started fining people heavily, and subsistence fishing became confined to a freshwater river and lake inland. Vasily is proud of having fought for years for a small-scale license (*lyubitel'skii uchastok*) for Naybá people to fish in the sea for their own needs, finally succeeding in 2021. We joined him with a relative one morning, checking the nets under the ice 50 meters off the shore. However, industrial-scale fishing still partially feeds the village from further away, as one of Vladimir's companies holds fishing licenses in tidal places in the Lena River Delta area. Vladimir and his wife sell that fish in Yakutsk, enlarge their profit with some more trading, and bring some of the income back to Naybá. The long distance, imposed by federal law, between the fishing spots and the reindeer herding operations artificially separates two Indigenous economies that otherwise would belong together in a livelihood which makes the best ecological use of multiple animal partnerships (Stammler 2010). Hence, most people live with their daily work and imaginative anticipations forcibly confined to reindeer horizons.

The principle of aboriginality or being First People has less legal weight in Russia than in other Arctic countries, even though Yakutia is among those regions where Indigenous rights are codified best within Russia (Fondahl 2014; Stammler and Ivanova 2016a; Ivanova and Stammler 2017). Though Indigenous privileges nationwide include free medicine and priority for helicopter places, neither Vladimir nor Vasily seems to be using indigeneity as a resource for their anticipation or for their agency in making their projects happen. Unlike many other communities even within Yakutia (Goncharov 2023), where there is a significant discourse of indigeneity, it was only in the school and the House of Culture that we were presented with a folkloristic version of Eveny identity. Unless we specifically asked, people seemed to be simply locals, and indeed, neither Vasily's offshore nor Vladimir's freshwater fishing license discriminated on grounds of ethnicity.

Vladimir's approach yields results far beyond anticipatory fantasies, as he harnesses any possible opportunity to provide jobs for village residents. His lobbying with the Ministry for Agriculture in both Yakutsk and Moscow has given him new vehicles and the status of being the only selective breeding enterprise (*plemennoe khozyaistvo*) for reindeer in the Bulunski District. He proudly points out that this gives him an additional government subsidy of 7 million rubles per year, money that he spends on raising his herders' salary to retain younger men (though they have to work harder to meet the federal requirements for reindeer fitness for these breeding-class reindeer).

The MUP also breeds herding dogs and supplies them to other herding units to refresh their blood (Muuttoranta, Majuri, and Stammler 2022), and Vladimir also acquired another subsidy to restore the polar fox fur farm. He openly admits that this gives no output other than a government cheque to pay one person's salary: "It creates one job, for free for us." By the time we arrived in 2021, of the seventy polar foxes only thirteen were left, the others having been eaten by stray dogs or wolves, but Vladimir was aiming for a government grant to re-stock the farm. Only when the one paid staff lost interest in the job in 2023 did Vladimir decide to shut down the project. Instead, he started promoting the signature ethnic Sakha livelihood of horse herding, unexpected in Naybá because the Eveny are traditionally closer to reindeer. But horses are useful to his reindeer herders as they can carry heavier loads than reindeer through the summer bogs. This subsidy creates a job for one herder. When half the horse herd became feral in 2019 in the mountains, Vladimir enterprisingly switched his team to occasional hunting of horse meat (Ivanova and Stammler 2025: 254–55).

The mountains of Naybá are home to the Kharaulakh population of the *chubuk*, the northernmost and most genetically interesting variant of the snow sheep (*Ovis nivicola*). Vladimir hosted a Moscow oligarch's trophy-hunting club and recruited paid local guides among his herders. The trophy hunters pointed out a large potential market for such hunting expeditions, especially from China. But Vladimir has shrewdly imposed a limit of ten sheep per year, to preserve the population for subsistence hunting by his own herders as well as for the future. He established a trophy-hunting club with a modern website in Russian and English,[19] which he named with the Eveny word for polar star, *osikat*. As of summer 2023 this operation was continuing successfully, when osikat hosted an oligarch from the US for trophy hunting. In 2020 Vladimir cooperated with a Moscow research institute to count the *chubuk* population. The herders showed the scientists exactly where they graze, but Vladimir complained, "these Moscow people expected us to drive them everywhere, and didn't understand that the *chubuk* needs to be approached on foot: this is why our herders know about the remotest grazing corners of the population. They should have been thankful that we even hosted them in the first place!" In the resulting publication (Dotsev et al. 2021), the local herders' input is not acknowledged.

Anthropologists are often used as a resource by the communities they study, and we became directly involved in Vladimir's plans. The first of these trophy-hunting tours yielded enough money to pay for four people to visit us in northern Finland in autumn 2019, to study state-of-the-art techniques of slaughtering, butchery, and finance. Speedily translating anticipation into action, Vladimir obtained funding from the Sakha Republic's Ministry of Agriculture, and on his way back from Finland bought a Russian copy of a Finnish slaughterhouse. In October 2021 our team from Finland was invited to the official opening of this modern slaughterhouse, back then the only one of its kind in the entire Sakha Republic. For the opening of the slaughterhouse, Vladimir asked us to bring specialists in reindeer slaughtering techniques from Finland to Naybá, to show how to operate this facility. For us, this was a rare opportunity to engage in reciprocal co-creation of knowledge. While the Naybá people learned about reindeer practice in Finland and how to adapt it to local conditions, we were able to observe the entire trajectory from anticipation to fulfilment, and to see this high-end slaughterhouse being realized in this remote settlement, in which it was the only building with running water (figure 2.3).

However, much initiative and enterprise were needed on the way: Vladimir got the money from the Sakha Ministry of Agriculture, plus an additional 6 million rubles for the transport from Chelyabinsk to Naybá. But the condition was that he must organize the transportation through one specific company nominated by the ministry. Just the transport from Yakutsk to Tiksi already cost a fortune. He could have got it free through his contacts, but this was not allowed. Moreover, the transport company brought it only as far as Tiksi. How would they get it from there to Naybá when there is absolutely no road? The Naybá people brought the slaughterhouse on trucks all the way across the rocks and through the bogs of the tundra. But then the weather became bad, and they could not unload in Naybá because there was no crane there. The drivers of the trucks had to be fed by Vladimir's sister Sargylana for three weeks. Vladimir waited until the weather improved, first made an ice road with an all-terrain vehicle with caterpillar tracks (*vezdekhod*), and then fetched a crane mounted

Figure 2.3. *Inauguration photo of the brand-new modular slaughterhouse in Naybá, 2021. © Aytalina Ivanova.*

on a truck and drove it to Naybá. Only after that could they unload the slaughterhouse. Then it was a huge effort to put it up. He and his sister's husband did it, with some local workers. They were very anxious that some small parts might be missing from the packages. Previous experience had thrown up such problems with a vegetable store that they had built with the supplies provided by Vladimir's friend Nikolay Alexeyev, the boss of YakutOptTorg, a major company responsible for supplies from central Russia to the Sakha Republic: when they ran out of screws, they had had to wait a month for more screws, so no work was done during that time. The same happened in late 2023 with the new YakutOptTorg warehouse in Tiksi, where missing parts had to be brought in by air after the season for river navigation was over.[20]

Vladimir persuaded his funders in the ministry that this slaughterhouse would be strategically located in the vicinity of the NSR, via which the product could be shipped out to European Russia (though, in fact, it could more easily reach Yakutsk). This is just one of many indications that, while much anticipatory rhetoric focuses on China, actual infrastructural development relates almost entirely to needs, links, and movements—and investment—within Russia. Both our main protagonists, Vladimir and Vasily, point to the disbalance in cargo traffic between the Tiksi/Naybá area and central Russia. Indeed, this was long ago baked into old Soviet terminology right across the Arctic, that while Indigenous herders out in the tundra or taiga produced the vital meat which fed the towns and cities of Russian miners, it was those cities which were called "points of supply" (*punkty obespecheniya*). Sending produce to cities used to rely on helicopters, which are now nonexistent or prohibitively expensive, but Vladimir anticipates that it will be cheap or free of charge to ship his production out of Naybá or Tiksi on the vessels that bring supplies to these Arctic settlements along the NSR and return largely empty. This anticipation is also fueled by Vladimir's friendship with the boss of YakutOptTorg, which has a vessel stationed in Murmansk. We dreamed together of how Naybá reindeer meat, fish, and cloudberries might one day reach the delicacy restaurants in St Petersburg via the NSR and Murmansk. While in 2021 this was still on the horizon of expectation, in 2022 that ship brought supplies from Astrakhan for the first time to the port town of Chersky via the NSR port Tiksi.[21] But this was before the autumn slaughter provided the meat, and anyway there would have been no onward contacts yet from Murmansk. So for the time being, Vladimir and his wife have started drying reindeer meat Finnish-style and selling Naybá cloudberries, shipped on the Lena River upstream to Yakutsk.

The administration head, Vasily Balanov, also contributed to this process: he got funding to accompany Vladimir and the herders to Finland, and, inspired by what he saw in villages in Lapland, he implemented several smaller projects to increase the well-being of people in his village. In 2021 he gave us a tour of what he had achieved so far. He and his team had built a Finnish-style meeting place on the shore overlooking Naybá bay, with a fireplace for villagers to hang out and cook meat, inspired by the *laavu* wooden shelters that the Finnish state maintains in beautiful locations. Vladimir, with his focus on business, said "What the hell, who needs that?" but people have started coming and spending time there.

Vasily also commissioned an outdoor sports place like the ones he had seen on his Finland trip, next to the House of Culture where he had established a satellite connection for the village, with an internet speed better than most people have in Tiksi. He says that more needs to happen now to increase the residents' quality of life, and they have spent some of their tiny municipal budget on refurbishing one of the old houses built from the remains of the homes that Finnish deportees in the 1940s had built in a nearby settlement. More derelict houses need to be torn down, and Vasily encourages residents to join voluntary tidying up campaigns to prepare for this. He says these are small steps in preparation for the big project of the Naybá seaport. Meanwhile, in a grand gesture of optimism, he has set up a big reflective entry sign to Naybá, for people who travel to the village on the winter ice road, to transmit a positive welcoming energy to visitors and show that they are coming to a vibrant community.

Ghosts of Infrastructures Yet to Come

It is clearly advantageous for China to ship goods through the Northeast Passage and along the Arctic Ocean coast. China may use Russian ports, even if other countries are currently not doing so. But China may develop technologies to manage with much less logistical support from settlements on the Russian coast. Some of our well-informed contacts in Yakutsk argue that all these expectations about the Northern Sea Route and development of Tiksi and Naybá are empty fantasies, and that Chinese interest in the NSR is falling as China plans to bring goods (and influence) to Europe not by ship, but by their rapidly expanding railway network.

The NSR is more a Russian Federal interest, and most of the large-scale infrastructure investment comes from the Russian state. Russia has decided to enlarge Tiksi and clean up the bay, and in addition to build a deep-water port in Naybá, which may eventually handle the bulk of the cargo turnover, becoming an important base for developing new deposits on the shelf of the Laptev Sea, as well as for new mining projects in the neighboring districts of Ust-Yana and Verkhoyansk. The federal government has a security and military interest in addition to economic rationales, and the port of Naybá has made it to one of the sixteen flagship projects under Kremlin priority in February 2024.[22] The regional government of the Sakha Republic has allocated land (*zemleotvody*) in Naybá, but unlike the federal government in Moscow is much more interested in the central districts around Yakutsk itself than the Republic's far-flung, sparsely inhabited Arctic regions.

Reviewing the levels of interest, from international, to Federal, to Republic, down to local communities, the activities and opinions of our friends in Tiksi and Naybá suggest that the nature of infrastructure anticipations, and the possibilities for their fulfilment, are strongly connected with possibilities for local participation.

Will community anticipations be the same as official ones? Because of local involvement, the vision behind this agency goes beyond the end-date of finite projects, the limitation of so many externally driven schemes, to give a better chance of long-term maintenance. These projects are not vulnerable in the same way to cynicism and horizons of disappointment because their actualization follows paths in local social structure (kinship networks, workplace comradeship), and in local culture (values of community cooperation, cosmological links between past and future, and ideas about action and consequences). This is a culture of mutual dependence and support (the "law of the tundra," *zakon tundry*), survival, self-sufficiency, and if necessary of defiance, evasion or co-opting of official powers. They accomplish, actualize, and fulfil their anticipations by skillful manipulation of the levers available to them, in a setting where the gulfs between administrative levels can be bridged by a personal visit to an office or a naked drink between people of unequal power in a shared *banya* (sauna). Naybá enthusiasts, in particular, have a huge practical, entrepreneurial spirit—in domains where they can genuinely have agency: persuading local authorities, plus their own technical skills and sheer physical energy when not being constrained.

Every infrastructure project already contains the potential for its own future ruination (Gupta 2018). Stuff that has lived and now died can take on a ghostly quality. Magnani and Magnani (this volume) use the term "ghost infrastructure" for a piece of huge infrastructure in Norway that never even came to life, a vacuum of unfulfilled anticipation.

Throughout the Arctic, especially in Indigenous communities, the idea of ghostliness is often more than a metaphor. Thus, the cultural response to industrial ruination is distinct, and different from, for example, the East German shrinking city studied by Ringel (2018). Just as an animal's footprint in the autumn snow remains traceable until the summer melt, so in Indigenous Siberian thought, any human movement on a path, or an occupation of a camp site leaves a residue of one's presence (Vitebsky 2005; Lavrillier 2020 calls this an "active imprint"). This may be experienced by those coming after as something ghostly. In the Eveny village of Topolinoe, Indigenous Asiatic residents are disturbed by ghosts of blond Russian prisoners from a nearby former GULag who cannot be exorcized by local healers (Ulturgasheva 2012); in the nearby village of Uchugei, local people hired a shaman from Yakutsk to appease the local spirits and exorcise the ghosts of their own GULag (apparently successfully, Stammler 2011); far to the west in Arctic Norway, ghosts of babies abandoned generations ago can be redeemed by naming and baptizing them (Nergård et al. 2019). Indigenous cultures emphasize premonitions and the fulfilment of prophetic dreams in waking life (Vitebsky 2005: 285–310). This suggests a particular cultural understanding of the continuing influence of the past on the present but also of fulfilment or its failure—and their connection to taking control of the future. In his *Christmas Carol*, Dickens shows us ghosts of the past, but he also alerts us to ghosts of the present and the future: Scrooge is redeemed by learning the lesson of "the ghost of Christmas yet to come."

Russia has a particular tendency to industrial ghosts because there is little need to clear up abandoned sites. If Tiksi is made less attractive because of abandoned junk, Naybá becomes more attractive because it offers a prospect of starting again in a clean place. The rusting junk in the bay of Tiksi is a ghost of the past which Vasily Burtsev tried to exorcise through a clean-up operation (cf. Broz 2024 for toxic junk as a source of spiritual misfortune), while the Magnanis' example is the ghost of an intention which will never be actualized. But there can be ghosts of future infrastructure which could conceivably come into being, as the February 1, 2024 Kremlin order for the port of Naybá indicates (see previous endnote). But regardless of its completion, this is a space of anticipation combined with the agency to imagine them and take steps towards fulfilment—or to redirect their energy to enable other plans. The anticipation of the deep water port in Naybá may in itself turn out to be only a ghost of an intention, but the structure of local involvement means that it is already opening up people's agency today.

Trajectories of Anticipation

Large infrastructure projects encourage anticipations of schools, village halls, and jobs. But any hope of local "benefit sharing" (Wilson 2019) is ultimately passive, using a begging bowl to make the best of the limitations to one's agency (for example, a community's powerlessness to reject a proposed mega-project). Some Indigenous politicians and lawmakers in Yakutia have themselves internalized this approach, which becomes evident in the expert literature (Burtseva et al. 2019; Burtseva et al. 2023), focusing largely on compensation for damage rather than on building a constructive dialogue (Stammler and Wilson 2006) between the parties who co-inhabit Indigenous lands. The perception of local and Indigenous remote residents as being passive recipients of compensation has a long history and remains the dominant principle of local as well as regional politics and legislation in Yakutia (Gordyachkova and Kalavriy 2022). The shift from passively waiting for benefit sharing to a local do-it-yourself approach involves a shift in what it is that gets anticipated, but also in the people who do the anticipating as they seize on unexpected openings to exercise agency and create expectation in quite different domains. Vasily, Vladimir, and their fellow villagers go much further, to increase the living standards of their communities through the work of their own hands. Rather than just receiving a share of the revenue of someone else's big project—like the Chinese projects which are anticipated in the city but rarely happen—they create a project of their own. This is what gives them their horizon of expectation rather than of frustration or resignation, a glass half-full rather than half-empty. The gym, the KSK, the village cinema, the extra employment of a fur-farm attendant or a horse herder, the Finnish slaughterhouse, the idea of exporting luxury foods to central Russia: all these have been triggered or facilitated by the possibility of the Naybá port. Even if Moscow (and Beijing?) never bring this into being, these smaller local projects will still take off.

Florian Stammler has carried out fieldwork in Siberia over thirty years prior to 2022, with Indigenous herders, extractive industry workers, elders, and youth, on which he has published extensively. His books include *Reindeer Nomads Meet the Market* (2005); *Young People, Wellbeing and Placemaking in the Arctic* (2022, ed. with Reetta Toivanen); and *The Benefits of the Cold and Domestication* (2025, ed. with Hiroki Takakura). He is Research Professor in Anthropology and founder and coordinator of the Anthropology team at the Arctic Centre, University of Lapland, Finland.

Aytalina Ivanova is a Sakha historian and legal scholar who has been specializing in fieldwork-based legal anthropology for the last twenty years. She has published on Indigenous Peoples' law, extractive industries, and impact assessment regulation, as well as co-authoring the chapter on Russian legal anthropology in the *Oxford Handbook Law and Anthropology* (2022). After employments at North-East Federal University (Yakutsk) and UiT the Arctic University of Norway, she is currently working at the Arctic Centre, University of Lapland, Finland.

Piers Vitebsky is a social anthropologist specializing in religion and environment among Indigenous peoples in the Siberian Arctic and Tribal India. His books include *Dialogues with the Dead: The Discussion of Mortality among the Sora of Eastern India* (1993); *Reindeer People: Living with Animals and Spirits in Siberia* (2005, winner of the Kiriyama Prize for Non-Fiction); and *Living without the Dead: Loss and Redemption in a Jungle Cosmos* (2017). From 1986-2016 he was Head of Anthropology and Russian Northern Studies at the Scott Polar Research Institute in the University of Cambridge.

Acknowledgements

We thank our research partners in Yakutia for their continued friendship and cooperation throughout the years, especially Vladimir and Ekaterina Alekseyevy and Vasily Balanov in Naybá and Vasily Burtsev in Tiksi. Field research for this chapter has been supported by the Research Council of Finland project # 342462, and data analysis by the SSHRC WAGE partnership project "Subsidies or Welfare."

Notes

1. In *Siberian Times* August 1, 2015, http://siberiantimes.com/business/casestudy/news/n0337-could-chinese-money-finally-bridge-the-lena-river-and-usher-economic-upsurge/, retrieved February 9, 2024.
2. For the sceptical view see the opinion of politologist Andrey Danilov, March 23, 2022 at regnum.ru/news/economy/3542121.html, retrieved February 9, 2024. For the official news of continued construction see reporting in the newspaper *Izvestiya*, at https://iz.ru/1602554/2023-11-09/most-cherez-reku-lenu-v-iakutii-dostroiat-do-2028-goda, retrieved February 9, 2024.

3. Northern Forum for Sustainable Development Yakutsk 2022, speech of the Chinese consul in Khabarovsk Jiang Xiaoyang, minute 50–55 of the video at https://vk.com/video-211653948_456239082, or through the programme at https://conference.tass.ru/events/iv-severnyj-forum-po-ustojchivomu-razvitiyu, retrieved February 7, 2024.
4. Power of Siberia 1, According to Vice-Prime Minister Novak on November 22, 2023, at https://lenta.ru/news/2023/11/22/power/?ysclid=lsbqderhtc903305469, retrieved February 9, 2024. Power of Siberia 2 failure as of December 12, 2023, according to https://meduza.io/feature/2023/12/12/zapusk-truboprovoda-sila-sibiri-2-mog-by-prevratit-rossiyu-v-krupneyshego-postavschika-gaza-v-kitay-no-proekt-nuzhen-moskve-silnee-chem-pekinu?ysclid=lsbq6qvehc144925449 retrieved February 9, 2024.
5. https://www.reuters.com/business/energy/a-property-sells-10-gas-producer-yatek-zhejiang-energy-567-mln-2022-01-19/ retrieved August 5, 2025, LPG delivery reported at https://oilcapital.ru/news/2023-06-22/yatek-vpervye-otpravila-sug-v-kitay-2963585?ysclid=lsbsqvutg1355574553, retrieved August 5, 2025.
6. https://www.interfax.ru/business/892100, retrieved February 9, 2024.
7. The Chinese investment in the bridge apparently cost three times less than the Russian side paid, https://dzen.ru/a/ZPt8_rZlbXapzg0V?ysclid=m8eonijnqk356847218, retrieved August 5, 2025.
8. According to Interfax, March 21, 2023 at https://www.interfax.ru/russia/892174, retrieved February 9, 2024.
9. For example by Putin during the opening ceremony of the third Belt and Road Forum in October 2023 https://paluba.media/news/81685, retrieved August 5, 2025.
10. https://www.kommersant.ru/doc/2720067?ysclid=lsbvkyxcy7765110610, retrieved August 5, 2025.
11. Sergei Vitte, 2019, at https://sakhaday.ru/news/ni-na-kakie-sereznye-proekty-kitaya-v-yakutii-rasschityvat-ne-stoit-istochnik?ysclid=lsbw13y9i3505178292, retrieved August 5, 2025.
12. Vice Prime Minister Sadovnikov announcing on March 17, 2025 that construction is "88 percent completed" https://paluba.media/news/81685, retrieved August 5, 2025. Chinese replacement of western technology at https://ria.ru/20230829/yakutiya-1892840640.html?ysclid=lsbwg78bm0171391564 retrieved February 9, 2024.
13. On year-round navigation on the NSR, and Chinese involvement as of November 2023, see https://thediplomat.com/2023/11/arctic-ambitions-chinas-engagement-with-the-northern-sea-route/, retrieved August 5, 2025.
14. Vasily Burtsev, personal communication, referring to official data, available at https://cleanarctic.ru/tpost/lt90ijtnf1-800-tonn-za-nedelyu-v-posyolke-tiksi-ust retrieved August 5, 2025.
15. E.g. forum discussion at https://vk.com/wall-74008979_13233?ysclid=lsbyqu2af4905966281, retrieved August 5, 2025.
16. Anton Lavrov and Anna Cherepanova "Severnyi Morskoi Shchit," June 21, 2021, in the *Izvestiya* newspaper, at https://iz.ru/1181509/anton-lavrov-anna-cherepanova/severnyi-morskoi-shchit-zavershaetsia-stroitelstvo-arkticheskii-zenitnoi-kreposti, retrieved February 9, 2024.

17. http://arcticway.info/polyarnyy-ekspress-vols-vdol-severnogo-morskogo-puti, retrieved February 9, 2024. For stagnation as of March 2025, see https://www.comnews.ru/content/238257/2025-03-14/2025-w11/1007/polyarnyy-ekspress-vstal-pauzu?ysclid=m8fvckad6t632798257, retrieved August 5 2025.
18. As of March 19, 2025, https://nsr.rosatom.ru/en/ retrieved 31 March 2025.
19. https://osikat.com/, retrieved August 5, 2025.
20. The missing screws are even mentioned in the local news, albeit only as a side-note to the success-story of another new building in Tiksi, at https://ulus.media/2023/10/09/zavershenie-stroitelstva-torgovo-logisticheskogo-czentra-v-tiksi-ozhidaetsya-k-20-oktyabrya/?ysclid=lsbz7uhsoj676971696, retrieved 5 August 2025.
21. According to a press release by the government of Yakutia on October 3, 2022, at https://sakhalife.ru/yakutopttorg-vpervye-zavez-produkty-i-strojmaterialy-v-arkticheskie-rajony-cherez-sevmorput/, retrieved 5 August 2025.
22. President Putin's order 200 of February 1, 2024, p. 3, available at http://www.kremlin.ru/acts/assignments/orders/73379, retrieved 5 August 2025. More detail in a news feature by Anna Shchetinina, at https://goarctic.ru/news/sdelany-ocherednye-shagi-na-puti-k-sozdaniyu-glubokovodnogo-porta-v-vostochnoy-chasti-sevmorputi-v-s/, retrieved 5 August 2025.

References

Adams, Ria Maria, and Teresa Komu. 2021. "Radically Ordinary Lives: Young Rural Stayers and the Ingredients of the Good Life in Finnish Lapland." *YOUNG* 30(4): 361–76.

Adams, Ria-Maria, Alla Bolotova, and Ville Alasalmi. 2023. "Liveability under Shrinkage: Initiatives in the 'Capital of Pessimism' in Finland." *European Planning Studies* 31(1): 212–29.

Adams, Ria-Maria, Lukas Allemann, and Veli-Pekka Tynkkynen. 2022. "Youth Wellbeing in 'Atomic Towns.'" In *Young People, Wellbeing and Placemaking in the Arctic*, ed. Florian Stammler and Reetta Toivanen, 222–40. London: Routledge.

Argounova-Low, Tatiana. 2004. "Diamonds: A Contested Symbol in the Republic of Sakha (Yakutia)." In *Properties of Culture—Culture as Property: Pathways to Reform in Post-Soviet Siberia*, ed. Erich Kasten, 257–65 Berlin: Dietrich Reimer Verlag.

Argounova-Low, Tatiana, and Mikhail Prisyazhnyi. 2016. "Biography of a Road: Past and Present of the Siberian Doroga Lena." *Development and Change* 47(2): 367–87.

Bernt, Matthias, Annegret Haase, Katrin Grossmann, Matthew Cocks, Chris Couch, Caterina Cortese, and Robert Krzysztofik. 2014. "How Does (n't) Urban Shrinkage Get onto the Agenda? Experiences from Leipzig, Liverpool, Genoa and Bytom." *International Journal of Urban and Regional Research* 38(5): 1749–66.

Bolotova, Alla, and Florian Stammler. 2010. "How the North Became Home. Attachment to Place among Industrial Migrants in Murmansk Region"

In *Migration in the Circumpolar North: Issues and Contexts*, ed. Lee Huskey and Chris Southcott, 193–220. Alberta: CCI Press.

Bolotova, Alla. 2022. "Leaving or Staying?: Youth Agency and the Liveability of Industrial Towns in the Russian Arctic." In *Young People, Wellbeing and Placemaking in the Arctic*, ed. Florian Stammler and Reetta Toivanen, 53–74. London and New York: Routledge.

Broz, Ludek. 2024. *Evil Spirits and Rocket Debris: In Search of Lost Souls in Siberia*. Oxford and New York: Berghahn.

Burtseva, Evdokia I., Anatoly Nikolaevich Sleptsov, and Anna Semenovna Fedorova. 2023. "Osvoyeniye Mineral'nykh Resursov Na Territoriyakh Zhiznedeyatel'nosti Korennykh Malochislennykh Narodov v Arkticheskoy Zone Yakutii: Etnologicheskaya Ekspertiza Proyektov [Development of Mineral Resources in the Territories of Indigenous Minorities in the Arctic Zone of Yakutia: Ethnological Expertise of Projects]." *Arktika IXXI Vek Gumanitarnye Nauki* 4(34): 46–67.

Burtseva, Evdokia I., Ivan Potravny, Violetta Gassiy, Anatoly Sleptsov, and Valery Velichenko. 2019. "Issues of Estimating and Compensating for Losses to Indigenous Peoples in the Conditions of Industrial Development of the Arctic." *Arctic: Ecology and Economy* 1(33): 34–49.

Donahoe, Brian, and Joachim Otto Habeck, eds. 2011. *Reconstructing the House of Culture: Community, Self, and the Makings of Culture in Russia and Beyond*. Oxford and New York: Berghahn Books.

Dotsev, Arsen V., et al. 2021. "Mitochondrial DNA Analysis Clarifies Taxonomic Status of the Northernmost Snow Sheep (*Ovis nivicola*) Population." *Life* 11(3): 252.

Fondahl, Gail. (2014) "Where Is Indigenous? Legal Productions of Indigenous Space in the Russian North." In *Nomadic and Indigenous Spaces: Productions and Cognitions*, ed. Judith Miggelbrink, Joachim Otto Habeck, Nuccio Mazzullo, and Peter Koch, 77–91. Aldershot: Ashgate.

Gansauer, Grete, Johanna Lilius, and Ria-Maria Adams. 2025. "Affective Politics and Neoliberal Subjectivities in 'Left behind' Places: Counter-Narrating Regional Decline within/from Finland's 'Capital of Pessimism.'" *Environment and Planning C* online first.

Goncharov, Nikolay Sergeevich. 2023. "Resursy Etnichnosti v Lokal'nykh Strategiyakh Osvoyeniya Prostranstva Severnoy Yakutii [Ethnicity Resources in Local Strategies for the Development of Northern Yakutia]" *Severo-Vostochnyi Gumanitarnyi Vestnik [North-East Humanitarian Herald]* 43(2): 24–42.

Gordyachkova, Ol'ga Vital'yevna, and Tat'yana Yur'yevna Kalavriy. 2022. "Promyshlennoye Osvoyeniye Arktiki VS Traditsionnyy Obraz Zhizni: Opyt Upravleniya Sotsial'no-Ekonomicheskim Razvitiyem v Respublike Sakha (Yakutiya) [Industrial Development versus Arctic Traditional Lifestyle: Experience of Managing Socio-Economic Development in the Republic of Sakha]." *Eko* 10(580): 129–47.

Grossmann, Katrin, Ria-Maria Adams, Alla Bolotova, Maria Gunko, Johanna Lilius, Sven Messerschmidt, and Leona Sandmann. 2024. "Self-Responsibilization, the Municipality, and the State: Peripheralization Shaping Local Initiatives in Shrinking Small Towns." *Cities* 150(105067): 1–10.

Gukov, Aleksandr. 2013. *Gorod Tiksi [The City of Tiksi].* Moscow: Printkom Publishers.

Gupta, Akhil. 2018. "The Future in Ruins: Thoughts on the Temporality of Infrastructure." In *The Promise of Infrastructure*, ed. Nikhil Anand, Akhil Gupta, and Hannah Appel. Durham: Duke University Press.

Haase, Annegret, Dieter Rink, and Katrin Grossmann. 2016. "Shrinking Cities in Post-Socialist Europe: What Can We Learn from Their Analysis for Theory Building Today?" *Geografiska Annaler: Series B, Human Geography* 98(4): 305–19.

Haikola, Simon, and Jonas Anshelm. 2017. "The Making of Mining Expectations: Mining Romanticism and Historical Memory in a Neoliberal Political Landscape." *Social & Cultural Geography* February: 1–30.

Heleniak, Timothy E. 2001. "Migration and Restructuring in Post-Soviet Russia." *Demokratizatsiia* 9(4): 531–49.

Heleniak, Timothy E. 2009. "The Role of Attachment to Place in Migration Decisions of the Population of the Russian North." *Polar Geography* 32(1–2): 31–60.

Hill, Fiona, and Clifford Gaddy. 2003. *The Siberian Curse: How Communist Planners Left Russia out in the Cold.* Washington DC: Brookings Institution.

Hobart, Mark. 1993. "Introduction: The Growth of Ignorance?" In *An Anthropological Critique of Development*, ed. by Mark Hobart, 1–30. London and New York: Routledge.

Humphrey, Caroline. 1998. *Marx Went Away but Karl Stayed Behind* (updated edition of *1983 Karl Mark Collective: Economy, Society and Religion in a Siberian Collective Farm*). Ann Arbor: University of Michigan Press.

Ivanova, Aytalina, and Florian M. Stammler. 2017. "Mnogoobraziye Upravlyayemosti Prirodnymi Resursami v Rossiyskoy Arktike [Diversity of Natural Resource Governance in the Russian Arctic]." *Sibirskiye Istoricheskiye Issledovaniya [Siberian Historical Research]* 5(4): 210–25.

Ivanova, Aytalina, and Gail Fondahl. 2020. "Legal Reform, Governance, and Security in the Russian Arctic." In *Routledge Handbook of Arctic Security*, ed. by Gunhild Hoogensen Gjørv, Marc Lanteigne, and Horatio Sam-Aggrey, 295–307. London: Routledge.

Ivanova, Aytalina, and Florian Stammler. 2025. "Laws of Domestication and Domesticating the Law in Yakutian Human–Animal Relations." In *The Benefits of the Cold and Domestication*, ed. by Florian Stammler and Hiroki Takakura, (Arctic Worlds, vol 7) 251–75. Abingdon and Oxon: Routledge.

Khlinovskaya-Rockhill, Elena, Lena Sidorova, and Piers Vitebsky. 2022. "Where Is the Nomadic Family? Rigid Laws and Flexible Tundra Life in Northeast Siberia." *Nomadic Peoples* 26: 4–32.

Konstantinov, Yulian. 2000. "Pre-Soviet Pasts of Reindeer-Herding Collectives: Ethnographies of Transition in Murmansk Region." *Acta Borealia* 17(2): 49–64.

Konstantinov, Yulian. 2002. "Soviet and Post-Soviet Reindeer-Herding Collectives: Transitional Slogans in Murmansk Region." In *People and the Land. Pathways to Reform in Post-Soviet Siberia*, ed. Erich Kasten, 171–86. Berlin: Reimer.

Lavrillier, Alexandra. 2020. "'Spirit-Charged' Humans in Siberia: Interrelations between the Notions of the Individual ('Spirit Charge' and 'Active Imprint') and (Ritual) Action." *Arctic Anthropology* 57(1): 72–99.

Leonov, Sergey Nikolaevich, and Elena Anatol'evna Zaostrovskikh. 2021. "Razvitiye Severnogo Morskogo Puti i Rost Aktivnosti knr v Arktike Kak Predposylki Usileniya Transportnogo Karkasa Dal'nego Vostoka [Development of the Northern Sea Route and the Growth of China's Activity in the Arctic as Prerequisites for Strengthening the Transport Framework of the Far East]." *Regionalistika* 8(2): 54–70.

Magnani, Natalia, and Matthew Magnani (this volume). "Economies of Ghost Infrastructures in Arctic Norway." Oxford and New York: Berghahn Books.

Muuttoranta, Kirsi, Karoliina Majuri, and Florian Stammler. 2022. "Koiraton Poromies Ei Siperiassa Pärjää" [A Dogless Reindeer Herder Won't Succeed in Siberia]. *Poromies: Suomen Poronjalostusyhdistyksen Ja Porotalousväeston Äänenkannattaja* [The Reindeer Herder: The Journal of the Finnish Reindeer Herders Association] 91(2): 29–31.

Nergård, Jens-Ivar, Solveig Nergård, Piers Vitebsky, and Sally Wolfe. 2019. "Eahpáraš—Nyfortolkning av Fortellingene om Fornektet Barndom [Eahpáraš—Reinterpreting Narratives of Childhood Denied]." In *Kulturen som Patient: Uvanlige Møter for Vanlige Folk [Culture as Patient: Extraordinary Encounters for Ordinary People]*, ed. Jens-Ivar Nergård and Piers Vitebsky. Oslo: Universitetsforlaget.

Nuykina, Elena. 2011. "Resettlement from the Russian North: An Analysis of State-Induced Relocation Policy." *Arctic Centre Reports* 55. ed. Florian Stammler. Rovaniemi: Arctic Centre, University of Lapland. https://core.ac.uk/display/29574444?recSetID=.

———. 2014. 'Making a Viable City'. Ph.D. thesis, geography, Vienna: University of Vienna. http://othes.univie.ac.at/34120/1/2014-08-27_1068697.pdf.

Orttung, Robert W., ed. 2016. *Sustaining Russia's Arctic Cities: Resource Politics, Migration, and Climate Change*. New York: Berghahn Books.

Richardson, Harry W., and Chang Woon Nam, eds. 2014. *Shrinking Cities: A Global Perspective.* Abingdon, Oxon, and New York: Routledge.

Ringel, Felix. 2018. *Back to the Post-Industrial Future: An ethnography of Germany's Fastest-Shrinking City.* Oxford and New York: Berghahn Books.

Sidortsov, Roman, Aytalina Ivanova, and Florian Stammler. 2016. "Localizing Governance of Systemic Risks: A Case Study of the Power of Siberia Pipeline in Russia." *Energy Research & Social Science* 16: 54–68.

Stammler, Florian. 2010. "Animal Diversity and Its Social Significance among Arctic Pastoralists." In *Good to Eat, Good to Live with: Nomads and Animals in Northern Eurasia and Africa*, ed. Florian Stammler and Hiroki Takakura, 215–43. Sendai: Center for Northeast Asian studies, Tohoku University.

———. 2011. "Dukhi Gor Vdol' Kolymskoy Trassy" [Mountain Spirits along the Kolyma Tract]. *Kistelen Kyys* 22–24(47–49): 20–22.

Stammler, Florian, and Emma Wilson. 2006. "Dialogue for Development: An Exploration of Relations between Oil and Gas Companies, Communities and the State." In *Sibirica. Special Issue on the Oil and Gas Industry, Local*

Communities and the State, ed. Emma Wilson and Florian Stammler 5(2): 1–42. Oxford and New York: Berghahn Books.

Stammler, Florian, and Elena Khlinovskaya-Rockhill. 2011. "Einmal, 'Erde' Und Zurück. Bevölkerungsbewegung in Russlands Hohem Norden [Once, 'Earth' and Back. Population Movement in Russia's Far North]." *Osteuropa Logbuch Arktis: Der Raum, die Interessen und das Recht* (2–3): 347–70.

Stammler, Florian, and Aytalina Ivanova. 2016a. "Resources, Rights and Communities: Extractive Mega-Projects and Local People in the Russian Arctic." *Europe-Asia Studies* 68(7): 1220–44.

Stammler, Florian, and Aytalina Ivanova. 2016b. "Confrontation, Coexistence or Co-Ignorance? Negotiating Human-Resource Relations in Two Russian Regions." *The Extractive Industries and Society. Special Section: Arctic Voices: Expectations, Narratives and the Realities of Living with Extractive Industries in the Far North,* ed Emma Wilson and Florian Stammler 3(1): 60–72.

Stammler, Florian, Aytalina Ivanova, and Lena Sidorova. 2017. "The Ethnography of Memory in East Siberia: Do Life Histories from the Arctic Coast Matter?" *Arctic Anthropology* 54(2): 1–23.

Thompson, Niobe. 2008. *Settlers on the Edge: Identity and Modernization on Russia's Arctic Frontier.* Vancouver: UBC Press.

Tichotsky, John. 2000. *Russia's Diamond Colony: The Republic of Sakha.* Amsterdam: Harwood Academic.

Ulturgasheva, Olga. 2012. *Narrating the Future in Siberia: Childhood, Adolescence and Autobiography among the Eveny*. Oxford and New York: Berghahn Books.

Vitebsky, Piers. 2000. *Coping with Distance: Social, Economic and Environmental Change in the Sakha Republic (Yakutia), Northeast Siberia*, Cambridge: Scott Polar Research Institute for Royal Geographical Society and Gilchrist Educational Trust.

———. 2005. *Reindeer People: Living with Animals and Spirits in Siberia.* Boston: Houghton Mifflin and London: Harper Collins.

Wilson, Emma. 2019. "What Is Benefit Sharing? Respecting Indigenous Rights and Addressing Inequities in Arctic Resource Projects." *Resources* 8(74).

CHAPTER 3

Economies of Ghost Infrastructures in Arctic Norway

Natalia Magnani and Matthew Magnani

Grøtsund Port outside the city of Tromsø began construction in the early 2010s, propelled by anticipation of the commerce and industry it would bring to the local area. With melting ice emerged hopes for an alternative passage for shipping across the Arctic between Europe and Asia, and a link to Russia's Northern Sea Route (see Introduction and especially Chapter 2), capturing the imagination of city planners, developers, and politicians. Thus, the Port of Grøtsund in northern Norway, nestled in the Arctic inlet of Tønsvika, in the small village of Tønsnes, has been steered by grand visions of becoming a thriving hub of global trade connection. Like many infrastructures of Arctic silk roads, Grøtsund is only partially built, waiting latent in the imagination for a possible boom in investment. Meanwhile, the port generates economies of anticipation, from the seeking of capital for business and industry that could conceivably use the infrastructure, to the making of Stone Age heritage, Sámi cultural tourism, identity, and spirituality that will shape identity in the region beyond the port's anticipatory life.

Tromsø Havn, or Tromsø Port, bought Grøtsund (the site of an old World War II German fort and state military site), from the Norwegian military in 2006, and developed it into an industrial port area in 2015 with the support of Tromsø Kommune, the municipality of Tromsø. While dreams of industrial activity and investment have not materialized, economies of anticipation surrounding the port's potential growth have taken on their own inertia: jobs are maintained for port authority officials and boards who develop strategic plans and seek long-term investment, loans and other investment capital are acquired, funding is generated for archaeological surveying, and a Sámi tourism business thrives in the area zoned for commercial activity abutting the port. While some economies of anticipation such as those of the port authorities depend on the idea that the harbor infrastructure will grow, others like the Sámi reindeer tourism operations seek to keep it contained within present boundaries. The latter prevent port expansion by expanding the cultural making of place, memory, and spiritual landscapes in relation to a tourism economy. This parallels processes of industrial capitalist growth in other parts of the world, showing that capital is created in contested zones both by those who encourage and those who reject development (Cross 2014). As capital is forged, so too are relations to place and memory: the interactions of diverse actors with the materiality and visions of Grøtsund Port, and with each other, activate and generate converging and conflicting memories of long-term relation to the land.

Concepts of ruination in anthropology have developed in studies of infrastructure, industry, and the failed promises of capitalism. Akhil Gupta (2018: 73) describes large-scale infrastructures existing in constant interplay between "renewal and ruination," while Jamie Cross (2014: 2) illustrates the "landscapes of ruination" left behind by a failed port project, which inscribe unrealized dreams of development. In the case of the semi-operational, semi-abandoned port of Grøtsund, haunted by apparitions of development-to-be, and actual ghosts, the question emerges not only of whether infrastructures serve a purpose beyond function and politics (*sensu* Harvey et al. 2017), but whether the ruination itself generates its own "purpose," its own economies. If anticipation has socio-economic impact (Greiner 2016), then material ruination creates particularly fertile conditions for such anticipation—for imagining what is possible and what can be revived.

Grøtsund is what we in this volume have termed a *ghost infrastructure* characteristic of so many dreams for transport connectivity across the Arctic, made ghostly by the anticipation of what it could be. Hauntings and specters appear repeatedly across the literature on megaprojects and unbuilt infrastructure, as project plans are raised, disappear for reasons that may be political, economic, or merely practical, and are then resurrected. Carse and Kneas (2019) describe such projects as "zombie" infrastructures continually oscillating "between closure and renewal." These hauntings are of a different nature than those of ruins, wherein abandoned buildings evoke nostalgia for a bygone past (Grandin 2009), a project that could have been but failed

(Davis 1990; Grandin 2009; Yarrow 2017), or fears of precarity in "progress" (Johnson 2013, 2014). At the same time, both ruins and unbuilt infrastructures may not only be haunted by the past but by ghosts of possible futures.

The haunting need not be passive. As Stammler, Ivanova, and Vitebsky (this volume) show, ghosts of unrealized infrastructures may stimulate people to action, yielding material possibilities for local livelihoods. Likewise, the promise (and threat) of an expanded Grøtsund Port is mobilized by regional officials to seek large-scale investments, and by local residents to support entrepreneurial activity as well as a deeper sense of culture and heritage. By examining the diverse economies conjured around a ghost port infrastructure in coastal northern Norway, as local officials and developers seek to harness visions of connectivity, trade, and expanding sea routes across the circumpolar North, we argue that unfinished projects create alternative spaces for agentive action and economies of anticipation. We show that the confluence of these different anticipations shape a temporal texture of ghostliness in unrealized infrastructure, as articulations of heritage, contemporary identity, and speculation about future development coalesce around the unbuilt.

The study is ethnographically grounded in the city of Tromsø and nearby port area of Tønsnes. It especially centers on three months of fieldwork in 2022, which in addition to conversations with port authorities and board members, and university archaeologists in Tromsø, involved an immersive period of volunteering at a Sámi reindeer tourism camp directly adjacent to the port, whose activities intersected with those of both the archaeologists and port authorities. The fieldwork took place in the broader political geography of Sápmi, the transborder Sámi Indigenous homeland, where the authors have worked for a decade on related questions of cultural resurgence and heritage (e.g., Magnani, Guttorm, and Magnani 2018; Magnani 2022; Magnani and Magnani 2022, Magnani et al. 2023). Administrative planning and aerial maps, as well as methodologies following material relations in the landscape, provided another layer for understanding the interrelations between diverse stakeholders surrounding the port's anticipated construction.

Apparitions of Development

Tor Arne, a conservative politician now on the Tromsø Port Board, described the visions of connectivity that supported the initial planning of the port at the turn of the century; there was a boom in the petroleum industry, burgeoning industrial development in the Russian North, as well as discussion of the port's potential link to a Northern Sea Route facilitating trade and logistical connection between Europe and Asia with the melting of Arctic sea ice. Likewise, the port director, Jørn-Even Hanssen, remembered the "strong beliefs" in Russian industrial developments in Yamal, in oil and gas, and in the Northern Sea Route, that bolstered support for the plans. All of these visions, along with the argument of the port's ideal location just outside the city center, and associated economic opportunities for local communities and regional development, were used to justify and secure loans and investments for construction. Whether or not the Northern Sea Route and Asian trade connections would come to fruition was not so important, Jørn-Even explained. They were one way to secure start-up investments, persuade Tromsø municipality to approve a 600 million kroner loan (later retracted due to the loss of the Yamal contract), and get major oil, gas, shipping, waste, and other companies to enter into cooperation and potential-for-rental agreements (later not followed through).

The amalgamation of visions was successful, and hundreds of millions of Kroners were invested in the endeavor by the Tromsø port authorities. Immediately following construction, however, the petroleum sector had a sudden decline, and geopolitical relations between Norway (a NATO member) and Russia drastically transformed with the Russia-–Crimea situation in 2014. The result is that the port that is meant to be a bustling center of economic activity, instead exists as a constellation of material apparitions that give the impression of development-to-come in an otherwise quiet stretch of pavement jutting out into the strait. Only 6 to 7 percent of the 1.5 million square meter reserved industrial area—"industriområde"—is developed, and the port itself resembles a lone parking lot. Instead of port logistics, anticipatory conjurings and the economies they generate have come to constitute the day-to-day operations of the port.

For those working with the Tromsø Port administration, infrastructural ghosts emerge as singular apparitions of development evoking promises of what could be. At the Tromsø port authorities' meeting room, with dark blue walls and windows overlooking the fjords surrounding the island city of Tromsø, the port director ran through PowerPoint slides often used for promotion, to show some of the one-off activities that have been happening at the port since its official opening—one time support for the transport and assembly of wind turbines headed to Kvaløya (the nearby island), an oil rig maintained for fifteen months at the harbor, giving the appearance of operations yet preventing any other activity, and some supersized chains for nearby fish farms.

Similarly, the sporadic apparition of American nuclear submarines at the harbor, which come to change crew and restock supplies, has brought politics and anticipatory economies to the infrastructure's otherwise ghostly presence. Increasing the visibility of submarines, and remilitarizing the harbor, is part of a broader geopolitical strategy in the region as tensions continue to escalate between Russia and NATO; in past years, there has been more movement of Russian Northern Fleet vessels in the Norwegian Sea, and Norway has hosted allied NATO forces for exercises in the port area. When the American submarines first appeared, some residents from Tromsø city and the immediate area took up protest, and the media attention that whirled around the port brought the otherwise unassuming infrastructure into public focus. Tor Arne describes the coastal surfacing of subs as the literal surfacing of politics—both locally in Tromsø and geopolitically between East and West, surrounding pre-existing military movements in the area: He explains:

> It's what the port does. It takes ships [...] people are very against it because they probably don't know that the same subs are moving just outside and their ships are changing crew in the open seas [...] So I think it's more that it might have *visually shown* some people that, oh no, this is actually happening, and that it's been happening for 20 years. *It's easier to see a sub when it's at the surface.*

Apparitions of industry—from one-off fish farm chains to intermittently docking military subs—bring political attention, occasional income, and long-term bureaucratic economies of strategy and vision making. Mostly, it is presentations of the port to government funders and private investors, and not actual harbor activity, that form a major locus of Grøtsund's operations. Infrastructural visions are set out in a new "Master Plan," packaged as PowerPoint presentations at monthly meetings, public events, and pitches to potential investors. In a constant effort to stave off ruination and produce visions of renewal (*sensu* Gupta 2018), the port authorities attempt to conjure spirits of development by bringing the port's possibilities to public consciousness. When the prime minister and other government officials visit, they remind them about the port as a viable location for investment. Working with the concept of a "snowball effect," they "knock on the doors" of investors to draw more regular operations. The idea is that if they establish businesses, and raise name recognition, more businesses will come. In these ways, a half-operational dock becomes a platform for creating new economies dependent on imaginaries of development and their perpetuation, rather than the industrial activities that initially legitimized construction.

Thus the dreams continue, whether or not the port authorities actually believe in them. In 2020 the authorities ordered a report from the Centre for High North Logistics to assess the potential of a Northern Sea Route connection for Grøtsund. The report concluded that the Route and its traffic are not yet developed enough to realize the potential of Grøtsund as a link to Northeast Asian markets, which depend on an ice-free passage for the shipment of frozen fish, meat, and cargo. When we spoke, the port director also dismissed the feasibility of an NSR connection. Instead, he emphasized a new focus for the port—"grønne maritime skiftet"—the green maritime shift characterized by renewable energy and the maritime economy. The harbor is envisioned to provide support to growing industries in fish farming, wind plants, hydrogen production, logistics, mechanics, and sea waste. This too, had not yet materialized at the time of the fieldwork, beyond the sporadic transport of fish farm equipment and wind turbines featured in promotional presentations. Yet evocations of global trade and connectivity continue to fervently combine with ventures in domestic "green" industries, bolstering possibilities for renewed construction, and maintaining other economies that have emerged around the port's potential.

Excavating Sea Routes

The purchase of the land in Tønsvika by the port authorities in 2006 from the Norwegian military stimulated a set of procedures involving cultural heritage management. This included a plan and budget for archaeological excavations (11.5 million kroners), which put a team of archaeologists from the university in contact with Tromsø Port. Resulting excavations, conversations, and presentations constituted a heritage economy that emerged in anticipation of the development of the port's industrial zone, entwining the area's archaeological remains with strategic visions for the harbor and Sámi memories of long-term inhabitance (discussed later in the chapter). This was not the economy of a "heritage industry" (Geismar 2015), or explicitly politically economic initiative (Franquesa 2013), but rather an anticipatory economy wherein archaeological narratives of long-term human presence at the port site were co-opted by local officials to seek further investments for development.

In their publications, the excavation leaders cite Stone Age tools and a winter settlement from 7000 BCE, which determined that the place had long been a strategic location at the intersection of major waterways. The port stands in the inlet where the Tønsvik river runs into the sea to meet the straits of Grøtsund and Kvalsund. The convergence made the area rich with seal, fish, elk, and reindeer (Gjerde and Skandfer 2018: 70). Military presence in the area since World War II prevented residential development, and therefore many of the archaeological sites were not visible from the surface (ibid.). Meanwhile, a hill once an island surrounded by sea water, also had least development priority by Tromsø Havn (and therefore no funding for excavations); registered archaeological sites where the water would have been now outlining the area, today located on the property of the Sámi reindeer camp. The island-turned-hill is where the oldest finds are anticipated, but funding for further excavation cannot be obtained until a revival of industrial activity and plans for further expansion.

While the archaeologists made the port infrastructure ghostly by imbuing the ground underneath with a deep history, port officials and politicians mobilized these ghosts for the construction's future. One of the team's head archaeologists, Marianne, describes how local politicians would bring their foreign government and embassy guests to the excavations. The archaeologists gave presentations to these visitors, describing Tønsnes as a crossroads of major sea routes in deeper time. These narratives resonated with Tromsø politicians and port board members for whom investments and loans for the port depended on promises of connectivity. Marianne explains:

> They realized that ahh, this is actually something we can use in our image building. This is old, we called it "Tønsnes midt i leia," Tønsnes in the middle of the sailing route. And this was perfect for them. It was the same [as their vision]. They wanted to show that this harbor is in the middle of everything, it is perfectly positioned and it always has been. So that was very easy for us to demonstrate and to present, because that was what the archaeological material showed across an enormously long time frame.

The idea of a historical Tønsnes at the crossroads of major sea routes was quickly mobilized by the port authorities as confirmation of the location's suitability for the contemporary port project. If the place drew people to meet thousands of years ago as a winter site at the intersection of transport waterways, then today it could also connect northern Norway to a Northern Sea Route, and trade between Europe and Asia. While ships and companies dock at Grøtsund only sporadically, the port authorities evoke the site's historical centrality in northern exchange (with long-term occupation and long-distance connection, the oldest and largest houses), to conjure activity and imaginaries in the present.

Besides official government and port authority visitors, local people and school groups came to see the excavations, bringing with them things that further deepened the making of memory at the site. Some older people revealed objects that they had found in their fields over the years. Marianne explained that "stray finds from potato fields" had made people aware of older settlements. "They have a sense of Stone Age," she said. "But through the excavations, they got more knowledge about the action and the time depth, and what these finds actually mean in terms of a long history in their region."

Through heritage making in the industrial zone, economies surrounding the anticipation and planning of the port had longer lasting impact and durability than subsequent harbor activities. Just as the submarines docking outside of Tromsø brought collective awareness to existing military movements, so too the archaeological excavations surfaced and generated memories of Tønsnes as a hub of settlement and transport at a water highway crossroads, and brought them to public consciousness. This entanglement of subsurface materials with contemporary geopolitical imaginings was not based on large-scale resource extraction as has been the case across the contemporary Arctic (e.g., Nuttall 2017), but rather small local economies that cultivated a sense of connection to the area in deeper time.

Zoning Memory

Between the time the area for the harbor was acquired from the military, and the official opening of the port in 2015, there was a process of zoning and planning of the industrial area: some sites were designated for "næringsvirksomhet," or commercial and business activity. In one of the business-zoned areas Lásse, a reindeer herder, had been feeding his herd when winter pastures became difficult. To cope with the environmental changes, he began to entertain visitors who would pay for the experience of helping with the feeding. Due to the zoning regulations, Lásse turned his activities into a tourism "business," and applied to officially rent the land from Tromsø Havn. As he, his family, and partner camp owners anticipate further port development that could potentially end their tourism livelihood, they reinforce the boundaries of their land use, nurturing a landscape of Sámi songs, spirits, and stories. This draws more tourists and Tromsø residents who come to connect with Sámi culture. The owners of the camp hope that the more they establish the site as a Sámi place, and the more jobs for locals that the enterprise creates, the more likely that port authorities will allow them to remain on the land.

Lásse's family had been moving their reindeer across Tønsvika since World War II, when they relocated from inland areas of Finnmark. Border closings between Norway and Sweden after the war stopped reindeer migrations from the Swedish side, leaving possibilities for those from Kautokeino, Máze, and other inland areas to move to new pastures. From the 1960s, Lásse's family started using a fenced site abutting the old military infrastructure at Grøtsund as an enclosure for slaughtering reindeer and preparing them to be moved to winter pastures. Due to plans for the port and industrial area, the municipality arranged to build the family a new reindeer fence for the activities 10 kilometers away. But the previous fence was still there, and about ten years ago Lásse stayed behind while the rest of his family moved their reindeer to graze and roam across the winter pastures, while he instead fed his animals enclosed at the summer pastures, in the shadows of the port-to-be. The winters were becoming more difficult with climatic changes, warmer with more freezing and thawing that would trap food underneath sheets of ice. With the changes in ice conditions, Lásse explained, the reindeer started coming to the feeding area themselves: "the reindeer were used to the place," "they know the roads," referring to the road that connects to the city. At the same time, Norwegian conservation laws were increasing predator numbers and thus reindeer deaths. Lásse recounts that the year he stayed behind in the summer pastures, feeding his herd in the protected area behind the port, he did not lose any animals. So, for the next three winters he continued.

While Lásse was there alone caring for the reindeer, the place began to draw curious travelers who would come to Tromsø to see northern lights and find out that just outside the city there were reindeer. They began stopping at Lásse's camp and helping with the feeding. As the number of visitors grew, Lásse realized that he could have tourists help with the labor of winter feeding and pay for the expense (each reindeer can eat two kilos of feed a day). A safari company asked if they could arrange tours, so Lásse and some friends set up a small *lávvu*, Sámi tent, and began having overnight guests for a "Sámi reindeer experience." Thus, Lásse's lone reindeer feeding in the darkness of the polar nights slowly transformed into a tourist operation. Gradually, he acquired a larger team of people—friends, family, local residents, and seasonal workers, trained more reindeer for sledding, built wooden hut structures for meals and coffee and a larger *lávvu* that could hold 100 people for Sámi stories, and changed the "reindeer experience" model from small overnight groups to multiple daily tours for 40 to 100 tourists. When the snow comes in October, until the end of the tourist season at the beginning of April, Lásse has 300 reindeer in a large, fenced field behind the port. Over the months that the first author (Natalia) volunteered at the camp, tourists came to feed the animals, go on sled rides (figure 3.1), see northern lights, try reindeer stew, and hear stories and joik (a form of Sámi musical expression) from a Sámi guide. When the tourist season ends, Lásse and his family move the reindeer back to the mountains.

Despite the success of his entrepreneurial venture, Lásse's ability to maintain a "business" as part of the industrial area of the port is uncertain. He has a renewable five-year contract to use the land for tourism, but at any point, he can be given one year's notice to clear the way for a larger scale "business" (i.e., maritime logistics and storage) that brings high revenue for the port. And so the label of "business" for his tourism venture needs to be constantly defended and reaffirmed. Documentary filmmakers, Youtubers, and photographers who visit the camp are eagerly accepted both for the camp's advertising and for their potential to bolster an image of the camp's presence and local significance. When Natalia asked Lásse if she could write about her experiences at the camp, he answered affirmatively, "maybe we can use it [the academic work] if they tell us we need to leave."

Another camp owner remembers a promotional video circulated by Tromsø Havn, showing the whole harbor area, including the reindeer camp property, completely built up with port industries alongside an upgraded road. Although the advertisement seemed to the owners and employees to be an embodiment of "ridiculous unrealistic dreams," it nevertheless drove them to make the camp as big as possible and to employ more people from the surrounding area. They hoped that this enlargement and assertion of presence would demonstrate the camp's contribution to the port's stated goals of community building and economic revival, and if necessary, potentiate local resistance to their displacement by the harbor's further development. Lásse explained one day around a fire in one of the wooden constructions of the camp that whenever there are plans for new infrastructure or industry, Sámi livelihoods are positioned as unproductive and

Figure 3.1. *A view of the "Sámi reindeer experience" from the back of a reindeer sled, 2022. © Natalia Magnani.*

not contributing to the local economy. He says, "when they want to have the land, build the industry, they often go after Sámi, people say that you're not [paying taxes]. … you're not producing anything. You don't have working places, you don't have anything …"

Lásse asserted that they, too, were creating an industry, one that creates *arbeidsplasser*, “work places,” rooted in the land: “Hopefully we could have it like a nature industry area [...] my dream one day would be that we could buy this land and build it like the industrial infrastructure [that] *we* want. Maybe someday it will feel safe to be here, to not have it in the back of your mind that maybe someday ...”

The constant desire for “safety” amid a sense of uncertainty was palpable during the tourist tours. After Natalia helped gather a crowd in a large *lávvu* (Sámi tent) for Lásse’s stories of Sámi culture and history, he explained to the group that industrial developments encroach on nature, which hurt another “industry,” motioning to the nearly 100 tourists lit by the campfire.

For tourism economies that emerge in anticipation of development and rely in some way on the infrastructures or opportunities generated by that development, the resulting entrepreneurial endeavor is often precarious (Bennike 2020). In this case, the reindeer camp as an economy of anticipation is reliant on both the infrastructure of the port (the fence, the zoning, and the widened road), and its lack of development beyond existing boundaries. In this case, the infrastructure is made ghostly by ambivalences and uncertainties about its reliance, as the herders and entrepreneurs can only speculate about plans for the port’s expansion, and whether their license to operate in its vicinity will continue.

On the one hand, the military and port infrastructure, and later business zoning and widened road, created material conditions—both economic and infrastructural—for that particular spot to be used for feeding the reindeer and associated tourist experiences. The commercial zoning of an industrial area around the port is one of the factors that encouraged the initial establishment of Lásse’s business, and continues to drive growth in the hopes that the expansion of jobs and tourism revenue associated with the camp will better secure its place as a key “business” of the industrial zone. The road leading to the reindeer camp and port was expanded in preparation for a container harbor that never materialized, before plans for cooperation with large oil companies and Yamal projects dissipated. Meanwhile the fence, partially maintained by the Port where it abuts the industrial zone, and partially by Lásse, protects the reindeer in the winter from predators, and the marshland birds and berries prized by local residents in the summer and fall. Moreover, the long-term presence of the military and then the port have kept housing developers from the site. Lásse explains that he and the port in many ways work together.

At the same time as it provides certain infrastructural and administrative affordances, the threat of the port's physical expansion fuels the tourist camp as an economy of anticipation. A Tromsø Port sign on the outside of the joint port/reindeer fence reads in Norwegian "KUN GODKJENT FERDSEL TILLATT," "only approved traffic allowed," reminding that the symbiosis between the port and reindeer camp is delicate (see figure 3.2). For two years there was an oil rig at the harbor that bathed northern light and reindeer experience visitors in blinking light and sound. Even today a flood light from the harbor appears each evening, constantly reminding of the port's presence. Not knowing whether or when Lásse's family may have to leave, all the *lávvu* and wooden buildings in the camp were built with the capacity to be taken down in a couple of days. Meanwhile, the camp owners are making new structures in case they have to move their activities to a nearby location, where the other reindeer fence for the family stands. With the war in Ukraine and western boycotts of Russian oil, Lásse fears there may be a resurgence of Norwegian oil production, leading to increased need for the wider industrial area of the harbor. Thus, the reindeer tourism economy in Tønsvika, through the simultaneous assertion of presence and local economic importance, and a transportability of physical structures in case the camp needs to move, operates in anticipation of port expansion.

Meanwhile, Sámi memory and place-making further nurtures the reindeer camp and its future. Lásse and his family recall memories, form new songs and stories, and create a site of Sámi culture for visitors and nearby residents in ways that affirm their presence in the area and ultimately support the viability of the tourist camp. Similar to the ways that the anticipation of infrastructural development has been shown to stimulate new economies of land use and border redrawings (Greiner 2016), so, too, the looming possibilities of port expansion stimulate the territorial inscription of cultural memory, reinforcing the boundaries and presence of the camp. Like the building of tourism structures in the area zoned for "commercial activity," memory, too, becomes confined on one side of the port fence, within the designated business zone.

The "Sámi reindeer experience" draws people who come to connect with Sámi culture, from tourists to local residents to the people who work there. At the time of the fieldwork, some of the young guides were from Sea Sámi families that had experienced the deepest assimilation pressures from Norway's Norwegianization history. They sought to reconnect with their Sámi roots as part of a wider cultural resurgence in the coastal region, and the telling of stories,

Figure 3.2. *A Tromsø Port sign reading "only approved traffic allowed" marks the barbed wire fence between the reindeer camp and the port, with a view of the harbor beyond, 2022. © Natalia Magnani.*

wearing of Sámi dress, and immersion in the Northern Sámi language that were part of daily life at the camp became an avenue for this reconnection. The ongoing Sámi cultural movement has its roots in the ethnopolitical awakenings of the 1960s and 1970s, which resonated with that era's global Indigenous movements (Bjørklund et al. 2000). Sámi activists and diverse community actors across Norway sought to reverse the linguistic and cultural loss that had occurred through Norwegianizing education, religious, and state institutions since the mid-19th century (Minde 2003). Today, the spirit of cultural awakening continues in the coastal areas especially, where many Sámi from northern settlements where people had entirely hidden their Sámi identity (see, for example, Eidheim 1971), now constitute one of the most rapidly growing demographics in Sámi politics and cultural resurgence.

While Natalia was volunteering at the camp, there were Northern Sámi speakers who would come and go, finding a place where they could be immersed in Sámi culture in a city area where Sáminess is otherwise less visible (despite having one of the highest Sámi populations in the country). Natalia, too, was drawn to the camp as a place where she could practice speaking the Northern Sámi language. The more that people travel to the reindeer camp for diverse experiences of Sámi culture, the more it becomes reinforced as a Sámi place. And although the land that Lásse uses for tourism is zoned as a "business," the Sámi memories that are forming their challenge boundaries between business and culture.

While Sámi cultural experiences for tourists and jobs for locals are part of conscious efforts by the reindeer camp's owners to position the enterprise as a "business" of the port's industrial area, the songs and stories of long-term presence have deeper resonance as they shape Lásse's memories of intergenerational relation to the land. The construction of the industrial area since the 2010s, and the year-to-year anticipation of expanded port development that could displace the reindeer camp, have shaped vivid memories of the life that Lásse and his family lived in Tønsvika leading up to the announcement and planning of the harbor. Lásse recounts:

> When I was a young kid, I remember this place, very cold, very dark [...] But we were together [...] with all my family, [...] marking [the] reindeer, slaughtering reindeer [...] It was the first memory of this place, and, of course, then this place has these memories and it is something that is inside you. It's like you have this place. And then it happened. It was in 2010 or something. And they wanted to build ...

Other memories in the landscape have emerged in dialogue with the archaeological excavations carried out in the planning of the port, all of which were done right on the reindeer camp site. During Natalia's time at the camp, Lásse would tell visitors that his family had been living in and migrating across Tønsvika for generations, and that the place has a deep human history. He recounted how archaeologists found 10,000-year-old remains there, indicating a winter gathering location at the crossroads of major waterways. Outside his public tours, Lásse would sometimes joke that they should retrieve the oldest objects from the Tromsø museum and bury them at the reindeer camp site, to stop construction in case of future industrial expansion.

One fall day when Lásse was out fixing the fence that surrounds the camp he began to joik–a Sámi form of musical expression often in remembrance of people and places. As he continued, he realized he was joiking the land, Tønsvika, and the memories of his family and reindeer who have been there working or migrating through. In the same Sámi history and culture hour during which he tells about the area's archaeological history in the large *lávvu*, Lásse performs this joik and asks visitors not to film: This joik belongs to this place, he explains, before joiking "Tønsvika": *Tønsvika de juoiggan dan dál […] álo leamaš johtin geaidnu* (Tønsvika, I'm joiking it now … It's always been a migratory route). When Natalia asked Lásse about the meaning of this joik, he explained that Tønsvika has always been a place to move through, and that today with the possibility of port expansion, it continues to be so.

To illustrate this point, he described how the site has been a place of migration between summer and winter pastures, and how infrastructures of state and private development, and negotiations with authorities, have shaped this movement. When Lásse's family first came to the area from Máze and Guovdageaidnu following World War II, the land around the harbor was owned by the military. In the 1960s there was a boom of residential construction in Tønsvika that disrupted the reindeer's migration route. The herd was stuck in the summer pastures and could not move come October. But then Lásse's family got help from their *verddet*, a Sámi designation for sedentary individuals or families with whom migrating Sámi would stay along their route, and with whom they were embedded in reciprocal social, economic, and labor relations. Lásse tells how the "old lady of the house" wrote a letter to the king, asking if they could use the site to move reindeer to winter pastures. The king, in response, ordered the building of a reindeer fence and transport of the herd by military boat. "That was the start of this place," Lásse explained, "and since then we have been using it to move the reindeer."

Lásse's experience of Tønsvika as a "place to move through" in anticipation of port expansion, is thus a continuity of Sámi economies shaped in negotiation with the affordances and constraints of local environments and state infrastructures. This interplay can be described as the entwining of "material dependencies" with infrastructure and unbuilt environments articulated by Schweitzer and Povoroznyuk (2022). The road connecting Tønsnes to major state routes facilitated reindeer slaughtering and associated labor but brought residential construction that disrupted the migration route. The military created obstacles but also safe passage for the reindeer. And today the port simultaneously threatens displacement while providing partial fence infrastructure, commercial zoning, and a wide road favorable for tour buses, driving both the reindeer camp's assertion of presence and their preparations for onward movement.

While the port currently has very little industrial activity, the expectation of future development in the zoned area has generated touristic livelihoods, shaping Sámi relations and memories of the land that further support these economies. As Lásse, his family, and the other workers of the camp anticipate being told to leave the area, they reinforce memories of intergenerational presence in its vicinity, create new joiks and stories grounded in the day-to-day activities of the site, and thus maintain the "Sámi reindeer experience" as an anticipatory economy of port infrastructure.

Spiriting the Land

In remembering Tønsvika as a Sámi landscape, and continuing economies anticipating port expansion, it is not only spirits of development that are conjured, but also Sámi spirits in dialogue with the environment, military and port infrastructure, and archaeological remains.

Behind the *lávvu* and wooden structures of Lásse's camp is an area of marshland and fields leading to the sea where reindeer pasture, abutted by the port fence. In the winter the snow blankets the land in a winter wonderland scene of reindeer pulling sleds over an ever-expansive stretch of white—a landscape of snow and northern lights that draws tourists to the region. During this time the marsh underneath is frozen, and the camp can receive visitors. As April draws near and the end of the tourist season in Tromsø approaches, the snow begins melting. This reveals the marshy brown field, feed, and reindeer droppings that reduce the appeal of the camp for tourists, making reindeer sledding more difficult and eventually impossible, and leaving no clean snow for reindeer to hydrate. With the protection of frozen ice gone, moving across the area becomes unpredictable. Deep ponds that pocket the camp reemerge and must be avoided, both in guiding tourists and in doing maintenance and fence repair work. It is at this time that the reindeer begin to be moved to the mountains. Negotiating weather conditions, the reindeer's needs, and the Tromsø tourism season, the exact end date of the reindeer tours is never certain, and grows less so as the summers appear to come earlier with climate change.

The physical entanglement of a spiritual and industrial landscape has from the beginning shaped the existence of the reindeer camp in anticipatory relation to the port. One of the Sámi guides working with Lásse at the tourism camp since its early days, Lone—a trained marine biologist and retired school teacher—recounts how when they first arrived at the site in the shadow of the port and began welcoming guests, she and Lásse's wife began having strange dreams about spirits living below the marsh. They dreamed that reindeer had fallen through and were running under the ice. The first *lávvu* built for the camp had a feeling of unrest and workers avoided staying there alone. It was also around that time that some reindeer fell through the ice, and a crane being used to put up one of the built structures started sinking into the marsh. That is why Lone began to wander farther from the camp into the area of German bunkers near the old World War II fort closer to the harbor. She started giving gifts to appease the spirits. Meanwhile, Lásse began to joik Tønsvika. As a result of these gestures, Lone says, the spirits became more benevolent, the sense of unrest dissipated, and no reindeer have been lost to the marsh since. All the while tourist numbers have grown, and the port has not expanded into the area of the reindeer camp. These days, Lone explains, the people of the camp work together with the spirits.

As tourist activities dwindled in March, Natalia set out with Lone to see the wider area beyond the main camp. Throughout the tourist season Natalia had heard from other workers that the field behind the camp was dotted with German bunkers, canons, and the remains of archaeological excavations. This was where Lone came when she was not guiding, especially in the summer, to do ceremonies. The military, port, and archaeological infrastructure, and Lone's ceremonies, were intimately intertwined.

With the rapidly melting ice, Natalia was instructed to put on rubber boots available for the workers. The wooden structures grew smaller behind them as Natalia followed Lone across the field, watching the large cracks crisscrossing the ice for signs of danger. Lone recounted that just the other spring she had fallen in up to her waist. *It is safe to walk where the reindeer walk*, she and others at the camp assured. Lone and Natalia approached a raised "island"—a hill at the edge of sight from the *lávvu* and wooden huts before the land drops off toward the sea, and the hill around which camp workers pull reindeer sleds and tourists. On the hill stands a German bunker, whose raised profile visible from the camp had first prompted Natalia's questions about the land beyond. "This is where the city [Tromsø] should be," Lone asserted, evoking the archaeologists' renderings of the site at the intersection of three waterways, its strategic location for the Germans, and the island in the center holding the richest and oldest archaeological deposits. As they stood on top of the bunker, Lone told how the camp and the wider area abutting the port had been central to her reconnection with her Sámi roots. "It's been a very sacred place to come and say thank you for bringing the reindeer here, and my Sámi culture back. I had to know who *I* was before I could tell the guests," she explains.

Lone's family has been in the northern regions for centuries—she was born in Vardø where her father's family is from, at the easternmost edge of Norway where spying on Russian subs and other military activity has become the norm (Higgins 2017). She grew up around Tromsø, on the lands of her mother's family. During her childhood she was only told proudly about her Danish last name and heritage. Her transformation into a young adult came on the heels of over a century of Norwegianization that reverberated across society; from institutions of education and religion to day-to-day life, differences were ruthlessly suppressed for the good of a unified nation-state. Lone started to dig beyond this history when she began working at the reindeer camp. She explains that in past decades Sámi consciousness and cultural reclamation have been rising, and people across the coastal regions of northern Norway are realizing that if they have generations of family in the area "a lot of it is Sámi, almost all of it." In her case, her grandmother's mother spoke Sámi. The clearer her roots became through her genealogical research, the more she was able to tell as she gathered tourists in the large *lávvu* for stories of Sámi history and culture.

When Lone recounts Sámi cultural history to visitors, she fuses accounts of Sámi practice with Nordic neoshamanism, which emphasizes the common roots of both Norse and Sámi spirituality as far back as 5,000 years ago, especially before Christianization, and the mixed heritage of different peoples in the region (Kraft, Fonneland, and Lewis 2015). She tells guests about the finds of the archaeologists described earlier in the chapter, citing traces of habitation in the area from some 11,000 years ago, before communities identified as Sámi. Lone further emphasizes that people in the coastal regions have interacted and blended their cultures over the centuries—through trade, travel and migration, and intermarriage. At the same time, she acknowledges histories of Sámi subjugation by Nordic populations, from the confiscation of Sámi drums and persecution of *noaidi* spiritual leaders by the Christian church, to boarding school assimilation and discriminatory land policy by the Norwegian state, which encouraged people to change their names and hide their identities. Lone's narrative reflects the convergence of a Sámi Indigenous movement that has been thriving and evolving since the 1970s, and a Nordic neoshamanism movement in which she has been active through her leadership of the Tromsø branch of the Shamanic Association and other gatherings and initiatives, and through her daily spiritual practice. The Sámi dress, or *gákti*, that she made for guiding tours at the reindeer camp likewise embodies this convergence of values and beliefs; Lone drew on existing skills from her design business where she creates garments inspired by old Norse religions—capes and dresses with shamanic and reindeer symbols—and research on *gákti* designs from the Sea Sámi localities of her ancestry.

Behind the hill with the German bunker the land ceases to be marsh, instead encasing more military remains—bunkers and canons—from World War II, next to sieves and structures from summer excavations in the last decade. One worker tells how in the early years following the war, people used to get trapped in snares that dotted the landscape. Decades later, the archaeological excavations that were done in preparation for the port's development, altered the reindeer's movements so that they now stay away from the area. Although the archaeological team endeavored to restore the ground to its previous state, mountains of dirt and archaeological equipment remained. The soil became too loose, and the reindeer no longer passed there.

Nearby the excavation remains, many of the old German bunkers have curious arrangements of reindeer antler, eagle droppings, and the bone remains of reindeer, birds, and fish. Lone explains that these spots are where she leaves offerings and does her ceremonies—likewise a blend of Nordic neoshamanism and Sámi religion (figure 3.3). Over the months of the midnight sun, long after the snow has melted and the reindeer and people are gone, the place becomes a sanctuary for marshbirds, eagles, and even sheep

Figure 3.3. *Lone atop one of the German bunkers she uses for her ceremonies, 2022. © Natalia Magnani.*

who are protected from dogs and predators by the enclosures of the reindeer camp and the port and military infrastructure. The eagles choose the bunkers as a vantage point to scope out prey abundant in the enclosure, including the occasional dying reindeer. They leave behind feathers that draw Lone to these spots, since the neoshamanism that she practices imbues the eagle with spiritual significance. Lone places reindeer antler on top of the bunkers, thanking the site for attracting the reindeer, and bringing back her Sámi culture.

The archaeology and spirituality of the landscape directly shape Lone's reconnections. In a YouTube video in which she shares her spiritual journey with a global neoshamanist network, she shows the fields behind the reindeer camp while joiking and wearing *gákti*, and then describes the place as an ancient landscape where archaeologists have found settlements over 10,000-years-old. "The land has shaped me," Lone explains. To tourists at the reindeer camp, she tells that being Sámi is not about genetics but about culture, and culture is shaped by nature. Even though many practices have disappeared, she adds, they can be reclaimed through engagement with places that hold stories. For Lone, it is the land behind the reindeer camp, abutting the port, which holds these stories.

Military and industrial activities at the port also intersect with Lone's spiritual practice. During one of her ceremonies some years before Natalia met her at the camp, Lone had an encounter with three military personnel from the adjacent harbor: "So I was sitting here singing and praying. And then I got this weird feeling, and I turn around and they were behind [me], 20 meters away with machine guns and full equipment." It turned out that this was the day the controversial American nuclear submarines were making a port call.

As for the reindeer camp, port infrastructure provides both affordances and constraints for Lone's activities, and by extension the anticipatory economy of cultural tourism at the site. It creates a secluded spot of spiritually significant fauna where Lone practices her ceremonies and reinforces her sense of Sámi identity, which she then recounts to guests at the camp. At the same time, she fears port expansion into her ceremonial areas. Already the unexpected military use of the port has interfered both with Lone's ceremonies and wider camp activities–from the closing of the road for American submarine calls, to military exercises that can be heard while tourists are feeding reindeer. Lone says, "if it doesn't get more than it is now, then it's okay, then we can co-exist."

Spiritual reconnections, entangled with marsh, military, port, and archaeological landscapes, support the continued presence of the reindeer camp. Lone's acts of ceremony and conjuring of Sámi spirits engender and reinforce Sámi cultural and spiritual relations at the site. They make the place a Sámi place, both for the people who work there or live nearby, and for the tourists who come and hear the land's Sámi stories. In these ways the repurposing of military, port, and archaeological infrastructure as a Sámi spiritual landscape becomes a way for tourism operators to represent and affirm the area as one of Sámi culture, further legitimating its use for cultural economies in the port's commercial zone amid threats of industrial expansion. These practices in anticipation of the port show that ghost infrastructures are not only ghostly in the uncertainty of their planning but also in the spiritual landscapes that they can generate.

Economies of Ghost Infrastructures

This chapter has followed a confluence of agencies and economies stimulated by the planning and anticipation of future port development. The physical port, in a latent operational state, creates an infrastructure from which spirits of renewal can be conjured. Economies of heritage, venture capital, tourism, and the associated spiritual and memory landscapes that emerge, activate, and reorder relations and memories of place. In the process, the memories and relations become entangled, creating a material landscape marked by economies of anticipation. Expectations of port development tie people's everyday movements at once to memories of intergenerational presence and a stone age past at the intersection of transport corridors, and to wider geopolitics. These associations affirm Indigenous connections to land and the long-term importance of the site.

The case suggests broader patterns of social and economic reverberation surrounding the ghost infrastructures of late capitalism. In the absence of regular industrial activity, acts of anticipation by local actors become the day-to-day operations of infrastructure. Failure to realize large-scale operations creates spaces for local agency in development projects, wherein the infrastructures and their promises are co-opted by entrepreneurs, politicians, and everyday residents to realize alternative economies and desires. Meanwhile, differential anticipations create a textured ghostliness of infrastructure marked by entangled temporalities, from the material construction and its imagined futures, to the surrounding landscape where anticipation yields past spirits to bear on the present.

Natalia Magnani is Assistant Professor at the Department of Anthropology and Climate Change Institute at the University of Maine. She leads the Norwegian Research Council project "Arctic Silk Road" based at UiT The Arctic University of Norway. She has published numerous articles on the Sámi regions, including in *Current Anthropology* and *American Anthropologist*, and has two books forthcoming—*Silent Knowledge: Reviving Sámi Practice* (University of British Columbia Press) and *The Craft of Belonging: Material Culture and Social Boundaries in Sápmi* with Matthew Magnani (University of Toronto Press).

Matthew Magnani is Assistant Professor of Anthropology and Climate Change at the University of Maine, and a researcher on the "Arctic Silk Road" project based at UiT The Arctic University of Norway. He has published dozens of articles on material culture past and present, in journals such as *American Antiquity* and *Current Anthropology*, and is the co-author of *The Craft of Belonging: Material Culture and Social Boundaries in Sápmi* with Natalia Magnani (University of Toronto Press).

Thank you to the tourism entrepreneurs and guides, and local scholars, politicians, and port officials, who gave generously of their time and connections to make this chapter possible. This research was funded by the Norwegian Research Council.

References

Bennike, Rune Bolding. 2020. "Himalayan Futures: Tourism and the Anticipation of Development." *Canadian Journal of Development Studies/Revue Canadienne d'études du Développement* 41(1): 110–26.

Bjørklund, Ivar, Terje Brantenberg, Harald Eidheim, John Albert Kalstad, and Dikka Storm. 2000. *Sápmi: Becoming a Nation*. Tromsø: Tromsø Museum.

Carse, Ashley, and David Kneas. 2019. "Unbuilt and unfinished: The temporalities of infrastructure." *Environment and Society* 10(1): 9–28.

Cross, Jamie. 2014. *Dream Zones: Anticipating Capitalism and Development in India*. London: Pluto Press.

Davis, Mike. 1990. *City of Quartz: Excavating the Future in Los Angeles*. London: Verso.

Eidheim, Harald. 1971. "When Ethnic Identity Is a Social Stigma." In *Aspects of the Lappish Minority Situation*, ed. Harald Eidheim, 50–67. Oslo: Universitetsforlaget.

Franquesa, Jaume. 2013. "On Keeping and Selling: The Political Economy of Heritage Making in Contemporary Spain." *Current Anthropology* 54(3): 346–69.

Geismar, Haidy. 2015. "Anthropology and Heritage Regimes." *Annual Review of Anthropology* 44(1): 71–85.

Gjerde, Jan Magne, Marianne Skandfer, and Hans Peter Blankholm. 2018. "Large Mesolithic House-Pits at Tønsnes, Coastal Northern Norway: Evidence of a Winter Aggregation Site?" In *The Early Economy and Settlement in Northern Europe. Pioneering, Resource Use, Coping with Change*, ed. Hans Peter Blankholm, 59–76. Sheffield: Equinox Publishing.

Grandin, Greg. 2009. *Fordlandia: The Rise and Fall of Henry Ford's Forgotten Jungle City*. New York: Metropolitan Books.

Greiner, Clemens. 2016. "Land-Use Change, Territorial Restructuring, and Economies of Anticipation in Dryland Kenya." *Journal of Eastern African Studies* 10(3): 530–47.

Gupta, Akhil. 2018. "The Future in Ruins: Thoughts on the Temporality of Infrastructure." In *The Promise of Infrastructure*, ed. Nikhil Anand, Akhil Gupta, and Hannah Appel, 62–79. Durham: Duke University Press.

Johnson, Andrew Alan. 2013. "Progress and Its Ruins: Ghosts, Migrants, and the Uncanny in Thailand." *Cultural Anthropology* 28(2): 299–319.

———. 2014. *Ghosts of the New City: Spirits, Urbanity, and the Ruins of Progress in Chiang Mai*. Honolulu: University of Hawaii Press.

Kraft, Siv E., Trude Fonneland, and James R. Lewis, eds. 2015. *Nordic Neoshamanisms*. New York: Palgrave Macmillan.

Harvey, Penelope, Casper Bruun Jensen, and Atsuro Morita, eds. 2017. *Infrastructures and Social Complexity: A Companion*. London: Routledge.

Higgins, Andrew. 2017. "On a Tiny Norwegian Island, America Keeps an Eye on Russia." *New York Times*, 13 June. Retrieved October 15, 2022 from https://www.nytimes.com/2017/06/13/world/europe/arctic-norway-russia-radar.html.

Magnani, Natalia. 2022. "Commemorating Continuity: Reconciling Representations in Sää'm Land." In *The Sámi World*, ed. Valkonen, Sanna, Áile Aikio, Saara Alakorva, and Sigga-Marja Magga. London: Routledge.

Magnani, Natalia, and Matthew Magnani. 2022. "Decolonizing Production: Healing, Belonging, and Social Change in Sápmi." *Current Anthropology* 63(4): 386–406.

Magnani, Matthew, Anni Guttorm, and Natalia Magnani. 2018. "Three-Dimensional Community-Based Heritage Management of Indigenous Museum Collections: Archaeological Ethnography, Revitalization, and Repatriation at Sámi Museum Siida." *Journal of Cultural Heritage* 31: 162–69.

Magnani, Matthew, Jelena Porsanger, Sami Laiti, Natalia Magnani, Anne May Olli, Paula Rauhala, Samuel Valkeapää, and Eric Hollinger. 2023. "Small collections remembered: Sámi material culture and community-based digitization at the Smithsonian Institution." *Museum Anthropology* 46(2): 92–105.

Minde, Henry. 2003. "Assimilation of the Sami: Implementation and Consequences." *Acta Borealia* 20(2): 121–46.

Nuttall, Mark. 2017. *Climate, Society and Subsurface Politics in Greenland: Under the great ice*. London: Routledge.

Schweitzer, Peter, and Olga Povoroznyuk. 2022. "Infrastructural Legacies and Post-Soviet Transformations in Northern Sakha (Yakutiya), Russia." *Journal of Environmental Policy & Planning* 24(3): 297–308.

Stammler, Florian, Aytalina Ivanova, and Piers Vitebsky (this volume). "Vast Spaces and Inconsistent Anticipations in the Northeast Siberian Arctic." New York and London: Berghahn Books.

Yarrow, Thomas. 2017. "Remains of the Future: Rethinking the Space and Time of Ruination through the Volta Resettlement Project, Ghana." *Cultural Anthropology* 32(4): 566–91.

Chapter 4

Polar Silk Worlding

Imagination, Infrastructure, and Anticipation in Chinese Arctic Tourism

Ria-Maria Adams and Mia M. Bennett

The Polar Silk Road, the moniker for a suite of investments from China in the Arctic, is typically discussed in infrastructural and economic terms (Tillman et al. 2018; see also Introduction). The focus has also largely been on investments in the Russian Arctic, particularly following Russia's full-scale invasion of Ukraine in 2022, rather than across the circumpolar North. Yet the concept of the Polar Silk Road can be broadened to include transformations taking place in the Arctic, which are both material and discursive, and human and non-human (Fraser, this volume, similarly demonstrates a Polar Silk Road taking shape through cultural and heritage initiatives). To explore this broader Polar Silk Road, or what we call the "Polar Silk World," the concept of "worlding" is useful. The term was first introduced by Spivak (1985) in her critique of a British officer's travels across India in 1815. She argues that the act of traversing colonized lands and assuming their "uninscribed" nature "generates the force to make the 'native' see himself as 'other'" (p. 254). Tourism worlding is a process of co-creation (Chimirri and Ren 2022), with both visitors and locals responsible for iteratively transforming destinations.

Worlding sheds light on how the Arctic is being remade—not solely in a physical and material sense but in ways that are imaginary and anticipatory, too. Different processes of "worlding" have taken place at the top of the planet throughout history. Perhaps the earliest example of the region's emergence on the cusp of external imaginaries and narratives was the voyage of Pytheas (320–306 BC). The Greek geographer was the first known European to sight sea ice and the midnight sun. A millennium and a half later, at the height of British exploration of the Arctic in the 1800s, exotic, fantastical literature incorporating the region as motif reached its apogee with the publication of Mary Shelley's *Frankenstein* in 1818.

Similar journeys and (re-)inscribings, this time not only literary but also visual, are now being carried out by the Chinese state and Chinese tourists. Already, ample scholarship has examined the geopolitics of the Chinese state's efforts to manage and govern the Arctic (Hong 2014; Bennett 2015; Liu 2018; Havnes 2020). Enriching this state-level focus, our work seeks to examine the more mundane and everyday practices of worlding at the level of tourists, paying attention to their imaginings of the region, the infrastructure they use and misuse, and the anticipation that was latent in Arctic communities following the COVID-19 pandemic as they waited to resurrect the suspended world of Chinese Arctic tourism. To support our analysis, other definitions of worlding also prove useful. Ong (2011: 1) examines worlding through the circulation of Asian urban models, noting that "cities have always been the principal sites for launching world-conjuring projects". As the Chinese state expands into new frontiers such as the Arctic, it is worth examining worlding projects from their peripheries, too. Ultimately, we aim to understand experiences in these liminal spaces, which, like Asian metropolises, are spectacular sites in their own right—and how they feed back into broader processes of Chinese worlding. Rather than critique the "art of being global," we might ask: What is the art of being "Arctic," or of being a part of the Polar Silk world?

As exploring Polar Silk worldings requires engaging with both its imaginary and infrastructural elements, we situate our work within the literature on the temporalities of infrastructure. Carse and Kneas (2019), calling for greater engagement with infrastructure and temporality, argue that "unbuilt and unfinished infrastructures can become the axes of social worlds and sites where temporalities are knotted and reworked in unpredictable ways" (p. 19). From this point of departure, we contend that certain tourism infrastructures in the Arctic have become anticipatory infrastructures, shifting from supporting actual flows of people to imagined ones. Much like Gupta (2018), we conceptualize infrastructures as processes rather than fixed things, which are characterized by multiple temporalities. Elsewhere in this volume, Stammler et al. illustrate how the different layers of anticipation and agency become visible through various infrastructure projects. They describe how international promises heighten anticipation of development, even though it often fails to materialize. At the same time, they highlight how smaller-scale entrepreneurial activities are more likely to be realized than national megaprojects. Our case study in Rovaniemi similarly demonstrates that local agency is conducive to developing infrastructure, yet it also shows how non-local forces can reshape and use it in unanticipated ways.

In recent centuries, Nordic, Soviet/Russian, and British worldings have been some of the most salient non-Indigenous representations of the Arctic. Many of these comprised mythmaking practices that were embedded within broader nation-building projects and carried out by individual explorers and colonists, who often pursued infrastructure development, too. In the 21st century, European forays into the north have been joined by Asian and particularly Chinese voyages and their associated imaginings of the region. The democratization and expansion of Arctic travel means that visits to the world's northern extremes are no longer being carried out by state-sponsored explorers and colonists but rather by millions of tourists each year. Their travels now form a widespread "communicative practice" connecting the Arctic to the rest of the world (Sukhorukov 2021). These encounters between different cultures, technologies, and systems of knowledge serve to "world" the Arctic. In effect, this could be considered "tourism worlding"—a term that, in the literature on tourism, is most prominent in relation to a location in the Arctic: East Greenland (Chimirri and Ren 2022). Whether along the wild, icy coast of the world's largest island or in the interior of Finnish Lapland, the Arctic world(s) that tourists conjure may be invisible or impenetrable to those lacking the ability—linguistic, cultural, logistical, or otherwise—to perceive them even when they are geographically and physically present within it. As Descola (2014: 273) offers:

> The material and immaterial objects of our environment do not stand in the heavens of eternal ideas ready to be captured, however imperfectly, by our faculties, nor are they mere social constructs giving shape and meaning to a raw material; they are just clusters of properties some of which we detect, some of which we ignore.

Given the relative ease of overlooking worlds right in front of us, deliberate and careful ethnography, participant observation, and thick description are necessary to identify and reflect upon the worlding practices tied up with Chinese tourism activities in the Arctic. Tourism, as an activity that combines both the spectacular and the mundane, offers a unique insight into diverse worlding practices.

In the Arctic, Asian and specifically Chinese tourism grew rapidly for half a decade until the COVID-19 pandemic erupted (field notes August 16, 2022). Like tourists from other middle- and high-income countries, large groups of Chinese visitors travelled north to experience winter attractions such as the northern lights, huskies, reindeer, snow, saunas, snowmobile tours, and visits with Santa Claus. A growing literature attends to this trend in Arctic tourism, with examinations of the overall picture of Chinese tourism in Arctic regions (Lee 2020), Chinese adventure tourism (Martin and Ren 2020), and Chinese tourists' views on Arctic tourism (Huang et al. 2017; Cai et al. 2020). The geopolitics and resources tied up with the sector, especially in light of the outsized role of the Chinese state in directing outbound tourism flows, have also spurred critique (Bennett and Iaquinto 2021). A study of Chinese tourism in Svalbard (Sokolíčková 2022), which our case study in Rovaniemi resembles, illustrates the multi-layered cultural differences embedded within the tourism industry, including differences in language, beliefs, and understandings of leisure, privacy, and nature, which often challenge cross-cultural communication with Chinese tourists.

Rovaniemi, the hub of Finnish Lapland, is a useful site for examining Chinese tourism worlding. Prior to the pandemic, national airline Finnair linked almost a dozen cities in Greater China directly to Helsinki, which offered easy onward connections to Rovaniemi. In 2022, only flights to Shanghai and Hong Kong remained, on massively reduced schedules and at far higher prices than before. Although pandemic-related border restrictions are now a thing of the past, Finnair still has not been able to effectively rebuild its network of routes across Eurasia due to the closure of Russian airspace to Western carriers. This has forced Finnair to take much longer routings across the perimeter of Russia to reach countries like China and Japan, meaning it no longer can claim to offer Europe's fastest connections to Asia.

During the pandemic, much of the infrastructure and many services developed to support the Polar Silk Road of tourism became moribund. At the Santa Claus Village in Rovaniemi, one of the epicenters of Chinese tourism in the Arctic, we observed brochures in Mandarin at Santa Claus Village gathering dust on the shelves in 2022 (figure 4.1). At an ornately decorated Chinese seafood restaurant in

Figure 4.1. *Brochures in Mandarin, Rovaniemi, 2021. © Mia M. Bennett.*

central Rovaniemi, European and American diners filled the tables, but Chinese customers were noticeably missing after pandemic travel restrictions were lifted—an unusual absence for a city that had grown accustomed to large Chinese tour groups filling its streets. While the infrastructure built to cater to Chinese tourists remained in place throughout the pandemic, Chinese tourists only gradually returned even as Chinese border restrictions began to ease in early 2023. In contrast, tourism from other parts of the world has recovered dramatically, with the 2024/25 season witnessing an unprecedented surge in overall incoming flights and tourist numbers. In 2023, Rovaniemi recorded a milestone of 1.2 million overnight visitors, marking a nearly 30 percent increase from 2022 as tourism rebounded from pandemic-related disruptions (CNN 2024).

Within the context of these shifts, this chapter explores Chinese Arctic tourism worlding in Finnish Lapland through three different lenses: infrastructure, imaginations, and anticipation. The first section, on infrastructure, considers the material transformations that have taken place in the region to support the growing numbers of Chinese tourists. The second, on imaginations, draws on interviews with tourism practitioners in Lapland to offer insights into how Chinese tourists imagine the Arctic, and how these expectations differ from the realities they encounter once there. Finally, the third section, "anticipation," examines the views of individuals in Rovaniemi who experienced the drop in visitors during the pandemic and then awaited the return of Chinese tourists, typically with a mixture of hope and disdain.

In sum, this chapter offers some reflections on how infrastructural imaginaries provide both avenues and constraints for realizing individual agency, whether by a Chinese tourist realizing their dream of experiencing northern worlds or by tourism practitioners running their own tourism companies. In this sense, we argue that Chinese Arctic tourism, as a form of worlding, involves encounters both generative and antagonistic between different groups of people and their hopes, fears, imaginations, and realities. Our contribution adds cultural nuance to critiques of Chinese activities in the Arctic, which often focus on more unidirectional flows of capital and extractive infrastructures—accounts which potentially risk exacerbating fears of Yellow Peril and 'Polar Orientalism' (Dodds and Nuttall 2016).

Research Methodology and Ethical Considerations

While our specific research into Chinese tourism in Finnish Lapland started in November 2021, the first author lived in Rovaniemi for six months before the pandemic during the winter tourism season and had encounters with Chinese tourists and guides, when they were still present in significant numbers. Over the past several years, the two authors have been conducting related research on topics such as lifestyle migrants working in the tourism sector (see also Adams 2023), transport and tourism infrastructures (e.g., Klinger 2023;; Sinanan et al. 2025) and the geopolitics of tourism (Bennett and Iaquinto 2023; Iaquinto et al. 2024), with the lead author's last fieldwork visit to Rovaniemi in February 2025. As a result, our insights are grounded in long-term, in-depth knowledge about the region garnered both before, during, and after the pandemic. This temporal dimension is important given the ways in which Covid-19 has affected tourism, particularly from China.

The main research for this chapter was conducted between November 2021 and August 2022 within the framework of ERC Advanced Grant project "Building Arctic Futures: Transport

Infrastructures and Sustainable Northern Communities—InfraNorth." During three field trips ranging in duration between two to six weeks, twelve semi-structured interviews with informants were conducted. Two follow-up trips, each lasting one to two weeks, were conducted in June 2024 and February 2025, during which informal conversations offered additional insights into ongoing developments. We, the two authors, have been doing research in Finnish Lapland since 2018, which enabled us to connect with previous interlocutors. These repeated returns to the field, as discussed by Howell and Talle (2012), enabled us to gain a deeper understanding of former participants and their engagements with tourism infrastructures.

The empirical material consists of interviews, focus group discussions, and detailed field diary entries based on periods of participant observation during our fieldwork. While in northern Finland, we actively participated in popular tourist activities: We chased the northern lights, stayed in cottages, Airbnbs, and hotels in and around Rovaniemi in different seasons, visited the Santa Claus Village and SantaPark, and frequented popular cafes, bars and restaurants. We established contacts with former and active tourist guides and connected with Chinese entrepreneurs living and working in Finnish Lapland. We also spoke to residents, whose views regarding their experiences with Chinese tourists in the region are included, too. Most of our informants were based in the capital of Rovaniemi and ranged in age from twenty-one to sixty-five years old. A major limitation of our research was our lack of Chinese language skills and the difficulty of engaging with Chinese tourists, who typically travelled in large groups with packed itineraries, leaving little opportunity to interview them during their holidays. At the same time, the lead author's fluency in Finnish enabled close interactions with many tourism practitioners in Rovaniemi.

We manually classified, coded, and interpreted our material following Bernard (2006), who emphasizes the value of finding and coding themes that appear in both field notes and interviews. In keeping with standard ethical practice, we refrain from mentioning participants' names and places of residence to protect their identities, which is all the more important when they share potentially negative views towards their communities. All potential participants were made aware that participation was voluntary and that their responses would be anonymous. In addition, we have been careful not to share sensitive interview content within the community to protect the research participants' anonymity and maintain their trust.

Rovaniemi: World of Tourism, Tourism Worldings

In Lapland, development of tourism infrastucture began in the 1920s with the construction of the road to Petsamo, Finland's key Arctic port until it was captured by the Soviet Union during World War II in 1944. The area attracted tourists, particularly wealthy Finns, thanks to its landscapes, salmon fishing, an Orthodox monastery, and the Skolt Sámi (Saari et al. 2020). Rovaniemi's airport was built in 1940, which also created a portal to the rest of the world. Following the end of the war, Finnish tourism companies started using Sámi elements to market Lapland. To satisfy tourists' expectations, the tourism industry started selling its own versions of Sámi culture. This problematic packaging of "authenticity" has caused a great deal of harm to Sámi people (Saari et al. 2020; Mazzullo 2022).

Yet it was not until 1984 when the decline of forestry in Finnish Lapland began to spur unemployment (Falk and Vieru 2017) that the idea of promoting Finnish Lapland as the "true" home of Santa Claus took hold (Herva et al. 2020). To stimulate national and international tourism, the region began to invest in transportation infrastructures centered on Rovaniemi, the regional capital located directly on the Arctic Circle with 63,500 inhabitants. As a result of public and private investments, tourism became one of the region's main sources of income and jobs (Falk and Vieru 2017).

Each November, Rovaniemi awaits the return of tourists. As the days shorten, the place begins to look drastically different—and not only because of the snow, whether it falls from the sky or blasts from cannons. Rovaniemi, home to Finland's third-busiest airport despite only being its 12th largest city, becomes reconnected to the world each winter by air. Similar seasonal reconnections can be found elsewhere in the Arctic, such as in Svalbard, to which increasing numbers of Chinese tourists travelled following targeted marketing of the place through various media channels (Sokolíčková 2022). Air traffic to Rovaniemi from European cities recovered quickly after the pandemic: in winter 2022/23, flights to Rovaniemi arrived from cities such as Paris, Milan, Bristol, London, Manchester, Dublin, Brussels, and Düsseldorf (Talvitie 2022). Yet arrivals from China via Helsinki have yet to return to pre-pandemic levels. In November 2021, though Chinese tourists had not arrived in significant numbers in close to two years, a billboard at the entrance to Rovaniemi Airport continued to advertise—only in Chinese—Tasogare, a high-end coffee brand (figure 4.2). Finnair, which used to focus on its connections to Asia, added a route from Rovaniemi to Tromsø in December 2023—one of

Figure 4.2. *Advertisement billboard, Rovaniemi, 2022. © Ria-Maria Adams.*

thirteen new routes across Europe that season, and twenty-four total (Degiampietro 2024). In winter 2024/25, Finnair announced that it would add 12 percent more seats compared to the previous season. Most of these seats, though, are accommodating European rather than Asian passengers.

Aviation has played a key role in repositioning Rovaniemi within global networks. The expansion of air routes to Rovaniemi has made Finnish Lapland easily accessible to tourists from Asia, leading the region to figure heavily in Asian Arctic tourism worldings. The transportation infrastructure has rendered the city's built environment and atmosphere more global and less Finnish by reorienting them to serve foreign tourists in addition to local residents. Ice and igloo hotels, souvenir shops, and snowmachines blowing artificial snow create a sense of a magical winter wonderland in the Santa Claus Village. All of this infrastructure remakes Rovaniemi according to the tourist gaze, blurring the distinction between authenticity and artifice. This may prompt the question: if tourists are such a dominant demographic in the city, especially in winter, to whose world does

Rovaniemi belong? If it is partly to the tourist, then is there a multiplicity of tourist worlds?

We argue that there are. One tourism world of particular interest with a distinct temporality merging rapid growth pre-pandemic followed by a sudden suspension in which both imagination and anticipation figure heavily is the Chinese world. By examining imagination, infrastructure, and anticipation in relation to Chinese tourism in Finnish Lapland, we consider the region's insertion into the wider "Polar Silk World."

Imagination

Like their counterparts from other countries, Chinese tourists seek an "authentic" Arctic experience. Experiences of authenticity are hard to define, and there has been a vast literature on what it means in relation to tourism since MacCannell (1973). Rather than authenticity existing "out there" as a pure and unchanged object, it arguably exists where "individuals feel themselves to be in touch with both a 'real' world and with their 'real' selves" (Handler and Saxton 1988: 243). In other words, authenticity is relational and achieved when a tourist's experience matches both their imagination of the place and their self-image. Therefore, authenticity exists as much in the imagination as in the world-out-there. It depends, as we will highlight, as much on showing the tourist the physical world in front of them as on making it come alive through storytelling and on what Handler and Saxton (1988) call "narrative coherence." Specific to Chinese tourist experiences, authenticity has been described as achieved "through an intricate waltz of gazing, experiencing and desiring to de-familiarize familiar spaces while playing up the imaginary" (Wee et al. 2021: 570). In the liminal spaces between the cerebral and the physical, tourism worldings are conjured and materialize.

The European Arctic, within which Finnish Lapland lies, is perhaps more familiar to Europeans who have had a longer history of northern encounters than Chinese.[1] The Arctic, and Finnish Lapland in particular, has featured as an "exotic and mystical fantasyland in European mindscapes" (Herva et al. 2020: 376), an imaginary world underpinned by the fabricated realities of Rovaniemi's Santa Claus and Christmas-centric tourism industry. The sector's perpetuation of this seemingly innocuous holiday imaginary has a damaging effect on communities in Finnish Lapland, as it is grounded in colonial ideas reproduced by the tourism industry (Herva et al. 2020). Moreover, representations of Sámi have changed considerably over time, from the image of "a backward minority" to the "romanticized children of the forest" (Mazzullo 2022: 197). A key signifier of the mythical winter fantasyland are husky safaris, even though there is no historical or cultural background for the use of huskies in Sámi or Finnish northern communities (Herva et al. 2020). One informant born and raised in Finnish Lapland expressed serious distaste for these inauthentic practices that had been imported from elsewhere. Activities such as husky tours exemplify the "Disneyfication" of the European Arctic in which completely imaginary traditions and practices conjure a "winter wonderland" that is, in effect, a wintry *non-place*, to use Marc Augé's (1995) concept – one that risks becoming increasingly difficult to distinguish from other tourist destinations across the region.

While Santa sits at the center of the European experience of Finnish Lapland—and, by token, the European and American "worlding" of the Finnish Arctic—the mythological character is relatively absent in Chinese tourism worldings. A comparison of "Finnish Lapland" on the US version of Google Images, for instance, includes numerous pictures of Santa Claus within the first few results. In contrast, a search for Finnish Lapland (拉普蘭區) on Baidu, the most popular Chinese search engine, contains no images of Santa Claus high in the results. This suggests that Chinese tourists may seek out different attractions when they come to Finnish Lapland, which demand different infrastructure. Santa Claus is sometimes described as the "world's strongest brand," reflected by his seeming omnipresence across various products (Hall 2008). Yet this omnipresence is clearly not global as revealed by his absence from Chinese imaginings of the Arctic. Instead, other phenomena like the aurora borealis represent a major draw to Finnish Lapland – one that may, in fact, be more universal given its appeal to both Asian and Western tourists. A woman working at a local tourism agency explained:

> The northern lights are the absolute highlight for all tourists. The Japanese started that trend. In the beginning we had more Japanese tourists but then more and more Chinese started coming. And Japanese do not like to be in the same place with the Chinese. So the Japanese tourist numbers declined once the Chinese discovered the place. The Japanese used to have their own tourism agency in town with their own Japanese employees.

The northern lights have become one of the main attractions and a primary reason to visit Finnish Lapland and the wider Arctic in winter for tourists from around the world (Herva et al. 2020; Adams 2023;

Sinanan et al. 2025). This novel phenomenon is partly traceable to the growing importance of visual media such as photography and videos, along with the "Instagram gaze" (Cassinger and Thelander 2020), which drives flows of people to specific destinations. One tour guide who has worked with Asian groups for four years remarked, "Pictures are extremely important. Everywhere you go—you take a picture. If you have been somewhere, you also have to show that you have been there." That goes for the aurora, too: "Pics or it didn't happen," as the trite Instagram-era saying goes. In fact, capturing the phenomenon in a photograph is arguably *more* important than seeing it with one's own eyes. One tour operator who had welcomed many Chinese tourists before the pandemic recounted, "In the beginning we went for the northern lights, then to a cabin with a sauna, and then stayed there until 3:00 am because then you can maybe see the northern lights during the sauna package. We didn't see the northern lights but I did a picture for them and you could see some northern lights in the picture." The longer exposure time and higher sensitivity of a camera compared to human eyes legitimized the experience of something which, in fact, remained confined to the imagination. Photographs make real and permanent what is in many ways an optical illusion (Bennett 2024).

Realizing that the imaginaries underlying Chinese Arctic worldings differ from European Arctic worldings, tourism practitioners create distinct infrastructures to enable them to be brought to life. The aforementioned tour operator who created a northern lights and sauna tour for their Chinese customers stressed, "You need to sell the whole package—that includes delicious food." Another young Chinese entrepreneur in Rovaniemi expressed, "Storytelling is an essential part of Chinese culture: They love to hear stories about their activities and this is something that the Finns and the Arctic tourism industry has not utilized fully." He added:

> They like cabins, saunas, hot jacuzzis. There is not much storytelling to the sauna from the Finns. There are no people guiding them what the process of the sauna is. Finns are not used to telling stories about the sauna. Finns just get there and that's it. Chinese pay big money, and they want the services to match the payment. Expectations are different to other European customers. They are service-oriented.

Since Chinese imaginaries of Finnish Lapland are, in some sense, a blank slate, it is up to local tour guides, practitioners, and marketing agencies to conjure them.

Emphasizing the importance of constructed narratives in influencing Chinese tourists' desire to visit Finland in the first place, the aforementioned guide noted that Finnish marketing in the past had proven very successful. Many of the efforts were carried out in partnership with major Chinese companies such as Alipay. The third-party mobile payment platform created a promotional video available in English filmed at Santa Claus Village to market its services (Bennett and Iaquinto 2021). Popular films have also inspired Chinese tourists to visit Rovaniemi. The same Chinese tourist guide described:

> There was a movie in China—*Let's Be in Love*, or something like that—which was shot in Lapland in 2011 or 2012. President Xi was here too, and he is high level. Chinese people like to go where other high-level people go. Santa Claus is not the main selling point. It's mainly about the northern lights. There is also snow in China. They like snow but it's not the major reason to come.

Indeed, a framed photo of President Xi sitting with Old Saint Nick hangs on the wall alongside photos of the icon with dozens of other dignitaries, all of whom gaze fixedly upon the tourists from around the world making their way down the hall to Santa's office (figure 4.3).

Figure 4.3. *Framed photo of Chinese President Xi Jinping with Santa Claus, Rovaniemi, 2021. © Mia M. Bennett.*

Once in Rovaniemi, Chinese tourists transform their impressions of the place into photographs, videos, and micro-blogs posted to social media and into stories shared in real-life conversations. These mundane activities complement more formal branding efforts. Together, all of these stories, which are grounded in both imagination and reality, give rise to Chinese Arctic tourism worldings.

Infrastructure

Infrastructure development is often driven by local actors, who wield sufficient agency to realize projects at a municipal scale. In this manner, practices of worlding the Arctic are driven by those who live within the region. Local actors strive to make the destination live up to its imagining, whether self- or externally imposed. Turning imaginings into physical realities inevitably requires infrastructure, which the first author witnessed in August 2022 in Rovaniemi. Yet infrastructure forms a key part of worlding when its development is undertaken to suit a specific group of people from the outside, whose visions and aspirations then imprint upon the landscapes.

As the long summer days drew to a close, the snow cannons in Rovaniemi's Santa Claus Village sat stored behind fences. Climate change may be delaying the start of snowfall, but tourists still begin arriving in November, meaning snow must be present. The snow cannons stand ready to ensure that the trees around the popular tourist destination glitter with white powder. Otherwise, Santa Claus Village—a massive revenue generator for Rovaniemi and the reason why so many families travel to the city from Europe and Asia—runs the risk of problematically snow-free images dominating social media platforms. In early winter 2018, for instance, when British tourists arrived and failed to find themselves in a winter wonderland, they nicknamed Finnish Lapland "Crapland" (Herva et al. 2020). For northern places that have been branded as rural, winter idylls—practically all of the Arctic—ensuring "snow security" is a matter of survival (Hall 2014). Snow cannons are thus a form of infrastructure used to make manifest snowy imaginaries of the Arctic, which are becoming all the more fleeting in the Anthropocene. Branded territories like Finnish Lapland are very much, then, "simultaneously real, imaginary and symbolic" (Nyseth 2009: 7). During winter 2024/25, the topic of snow was frequently discussed on social media platforms such as Facebook groups, where tourists expressed concerns about there being too little snow and missing out on the "full winter experience."

While snow can be manufactured, the aurora cannot. This vexing ephemerality, however, has not stopped local actors from pursuing infrastructural changes to make the ionospheric phenomenon easier to see, or, more importantly, capture on film. A significant amount of tourism infrastructure in Rovaniemi has been erected to deliver the northern lights to tourists, especially those from Asia. Many local tourism operators hold, rightly or wrongly, that tourists from Japan and China believe that children conceived under the aurora will be lucky, or that, for the Japanese, consummating a marriage under the northern lights will ensure a long and fruitful life (Cole 2018). These rumors, many of which are unfounded, have propelled infrastructural changes that physically reshape tourism infrastructure to offer what practitioners *imagine* visitors from Asia might want, even if this is not exactly the case.

A primary illustration of shifts to tourism infrastructure in Finnish Lapland are glass igloos, which are being built not just in Finnish Lapland but across Arctic communities. These structures accommodate tourists' desire to watch the aurora all night long from a warm and comfortable place. One informant, a young tour guide who worked with Chinese groups, explained, "They really like cottages and these glass igloos, especially if the cottages have saunas and jacuzzis." Such infrastructure offers a chance to view the aurora comfortably – yet at a steep price of over €800 a night for a two-person igloo in the high season. While, as our observations showed, some glass igloos can be in poorer condition than cabins, since they are perceived as offering an "authentic" experience, many Asian tourists splurge on a stay, even if only for one night.

Before the pandemic, another tour operator offered a creative infrastructural adaptation to bring the northern lights to life: a chartered passenger plane. The "northern lights flight" would fly above the clouds, and only window seats would be sold, guaranteeing all passengers a glimpse of the aurora. Although these kinds of chartered passenger aircraft flights are exceptionally rare, this instance demonstrates how far—or high—travel entrepreneurs and tourists alike are willing to go to guarantee glimpsing the northern lights. Ironically, sometimes tourists on their way to the Arctic don't even notice the aurora are outside the window of their airplane, as the second author noticed on a flight from Seattle to Reykjavik in February 2025. The northern lights were dancing outside the window over the Canadian Arctic, but most passengers—including one who, upon disembarking, told the pilot she had come to Iceland with hopes of seeing the northern lights, which she had never before seen—were asleep or unaware.

In effect, some tourists have come to perceive the aurora as an infrastructure like any other that is always present in their Arctic destination, running smoothly and seamlessly overhead, at least in winter. The reality is far from the case. A Chinese tour guide working in Rovaniemi offered a rather extreme example of when a Chinese tourist asked him to "start the show." The tourist expected that the guide could press some kind of "button" for visitors to see and to photograph the phenomenon.

While these examples may seem exaggerated, they demonstrate how little some tourists know about the northern lights and how much pressure there is on tour guides to turn the imaginary of the Arctic as bathed in the iridescent glow of the aurora into a round-the-clock infrastructure. One Chinese tour guide who had worked in the tourism industry in Rovaniemi explained how daytime activities were easier to handle because people usually get to do what they have been promised: ice fishing, visiting a reindeer farm, husky rides, snowshoeing, skiing, and so on. In contrast, it was harder to satisfy customers' wishes with nighttime activities, especially seeing the northern lights, which depends on weather, solar activity, and sheer luck.

Changes to the rhythms and speed of tourism infrastructure are also necessary to realize the fast-paced temporality of Chinese Arctic worldings. As one former Chinese tour guide based in Rovaniemi indicated, "Chinese tourists do not want to do the same activity twice. They also do not want to stay a long time in one place. If a European tourist would do two activities a day, the Chinese would book five activities on a day." A preference for a faster pace combined with the generally fewer days of vacation that Chinese have compared to Europeans results in more condensed tours and, ultimately, a faster-paced Arctic world. A desire for numerous activities a day condensed into a short visit has also helped to generate relatively high margins for tourism operators serving Chinese visitors (field notes, August 8, 2022). Moreover, one operator noted that Asian tourists are willing to book aurora tours multiple times. She noted, "Asian tourists would book up to three 'northern lights tours' in a row to increase their chances of actually seeing the nature spectacle. One can only imagine the disappointment if the aurora borealis doesn't show due to cloudy weather." She voiced concerns about how stressful it is when visitors end up disappointed or even angry if they are unable to see the ionospheric spectacle.

Separately, a Chinese entrepreneur explained:

> If they don't see the northern lights, the trip still needs to be *meaningful* for them. So, you need to come up with some other services. Nice food,

> sauna, or stories. Chinese clients can be very unreasonable. They are willing to pay a lot of money if you can show them the northern lights. So, depending on how much they pay, that's how far you are willing to drive to try to find clear skies. For example, to Levi or all the way to Sweden or Norway.

Again, the importance of storytelling and Arctic worlding becomes clear. Local agents must anticipate what their tourists want and then develop the infrastructure necessary to realize their dreams.

In the meantime, while these activities are generating memories and sparking the imagination of Chinese tourists, they are filling the wallets of local tourism operators. One local taxi driver reported of a customer paying up to €500 to reach a place where they thought northern lights might be visible. The former tour guide added that tourist operators will still run northern lights tours even if they know it will be cloudy and overcast, with poor visibility. "Because there is a sales opportunity," he said. "Northern lights tours have the highest margins. There is no cost but the fuel. In one night, you can earn more than €500 just to bring a group to the nature." Such efforts can represent a misuse or even abuse of tourists' time. These more negative encounters, however, go both ways. Just as operators can mistreat tourists, tourists can misuse local infrastructure, as we next explain.

(Mis)using Local Infrastructure: Effects on Local Communities

Tourists do not only rely on infrastructure built for them. They also use buildings, transportation, and services that were constructed to support local activities. This can put pressure on residents, as seen by the widely documented effects of Airbnb causing rents to rise, which several informants noted occurring in Rovaniemi. When new air routes are opened, apartment rental prices rise as it becomes more lucrative for landlords to offer short-term accommodation to the growing number of visitors than long-term leases to residents.

Visitors also often do not use local infrastructure in the same way as residents, which can cause friction and potentially even fatalities. Roads can become hazardous during the seasonal rush to see Santa. One resident described how he doesn't always feel safe driving on popular tourist routes in winter, as some of the rental cars are driven by visitors who don't know how to properly drive in Arctic winter conditions. In winter 2024/25, when the number of tourists reached an all-time high, a guide from the Finnish Road Safety Administration (*Liikenneturva*) was circulated on social media platforms with instructions for tourists on how to drive in winter conditions, with a specific section titled, "Don't become a hazard when sightseeing." It included tips such as: "When looking for the northern lights, reindeer or sights, think about your safety and the safety of others," and "Don't stop your vehicle in the middle of the road" (Finnish Road Safety Administration 2025). Locals also reported that tourists often drive dangerously slowly on the main roads, causing delays when trying to take pictures or simply because they are afraid to drive at normal speeds in Arctic road conditions.

Lodgings can also be used inappropriately. One recently retired couple who had started renting out the upper floor of their home recalled their first encounter with Chinese tourists:

> We have a small balcony upstairs and our Chinese guests put out the camera on a stand outside for the entire night, so that the camera could take automatically pictures in frequent intervals in case the northern lights would show. For the camera not to freeze, the balcony door was left open for most of the night, and the room temperature was very cold after some time. We put up signs to our flat indicating in Chinese that the doors are not meant to be open when temperatures are below freezing.

Here, Chinese tourists misused existing infrastructure—a house in Rovaniemi—in an attempt to realize their hope of seeing the aurora, or at least having their camera capture it. This occurrence shows how cultural norms and understandings surrounding the appropriate use of infrastructure can vary significantly.

Beyond residential infrastructure, transportation infrastructure can be misused, too. One Rovaniemi resident expressed:

> The Chinese tourists are a very visible group. You cannot not see them when they arrive with their huge suitcases. They were walking on the railroads and on the motorway towards town and crossing roads where you really better not cross a road … I don't know what it is, but maybe they have not seen snow because they sometimes do not use the sidewalks but walk in the middle of the driving road. Interesting because they come from cities where millions of people live and they should know not to walk on the road.

Chinese tourists have also been known to hire cars only to drive on cross-country ski trails or end up in snowdrifts, keeping tow-truck companies busy in winter. The preference to stay in a car rather than be out in nature may emerge, as one tour operator suggested, from a desire to avoid the cold. One Chinese informant indicated, "Chinese people are not fond of hiking. But skiing has started to grow in popularity. Winter sports is getting more interesting, but generally Chinese people don't like the cold, and they are not properly dressed." A tour operator echoed this sentiment, underscoring that although most of the activities in Rovaniemi, even in winter, are centered on the outdoors, "They are not properly dressed for the weather—we need to provide outdoor clothing for them." Similarly in Svalbard, Chinese tourists were sighted wearing inappropriate winter clothing and asking for "North Pole tours" while wearing sneakers (Sokolíčková 2022). A Chinese tour operator in Rovaniemi elaborated on his perceptions of working with Chinese tourists:

> You have to understand, they are not really outdoor people. They love seeing and taking pictures *of* the nature but they do not want to be outdoors *in* the cold for a long time. If you stay somewhere too long, they will ask you when we continue to the next place. And honestly, there is not that much to show them here in Rovaniemi. There is the forest, the Santa Claus Village, a few restaurants, the bridge across the river and—do not get me wrong, but the city center of Rovaniemi is not exactly a beauty.

Anticipation

While Chinese tourism reshaped Rovaniemi in the years leading up to the pandemic, that world was quickly suspended. Although European tourism to Santa's hometown rebounded to the point of exceeding pre-pandemic levels (Talvitie 2022), the Chinese government's continued border restrictions, including stiff quarantine measures for returning travelers, led the country's total outbound tourism numbers to drop from 155 million in 2019 to 8.5 million in 2021 (Bhutia 2022). While Chinese tourism in Finland still has not returned to pre-pandemic levels, the infrastructure built to support their visits remains, as do many of the members of the Chinese diaspora who had migrated to Rovaniemi to support and develop the previously burgeoning sector.

During the pandemic, tourism infrastructure reverted from supporting flows of people to becoming an anticipatory infrastructure, one in which Chinese tourists formed a "present absence" (cf. Carse and Kneas 2019). Rather than being developed in the spirit of, "If you build it, they will come," like so many ill-fated Arctic megaprojects, the operators of Chinese tourism infrastructure in Rovaniemi anticipate and even expect their former users to return. Both locals and tour guides, however, noted that in winter 2024/25, South Asians, particularly people from India, were becoming a more visible Asian tourist presence in Finnish Lapland, with many making use of the infrastructure previously designed to support Chinese tourists.

Whereas Chinese imaginaries "re-worlded" Rovaniemi, residents wonder whether the city's future will involve Chinese tourists, and if so, what that will look like. Just after the pandemic, one local entrepreneur offered, "Asia is an interesting market because there are so many people, but at the moment, we are focusing more on the European markets because nobody knows when and how the Chinese tourists will return." Similarly in 2022, another tourist entrepreneur remained pessimistic about the outlook for Chinese tourism in the Arctic. He expressed, "There is still no hope for the foreseeable future to get Asian tourists back. Only Asians who already live in Europe would come to Arctic destinations. However, they are very few individuals compared to the large numbers of tourists who started to come before the pandemic." Furthermore, pointing to the ambivalence with which Chinese tourism is viewed in the Arctic and the ways in which it is colored by undertones of Yellow Peril, the informant added, "This place will not be the same again when the Chinese return. I think people do not understand how many Chinese there are out there, just waiting to travel again and once they start coming again, they never come alone." Yet with the lack of easy air routes from China to Finland, the logistics of Chinese travel to Rovaniemi are now more complicated. After the pandemic, one interlocutor noted, "It will be interesting to see if they will return. With no flights possible over Russia, Finnair suffers." For now, in the place of Chinese tourists are tourists from other origins both near and far. One tour guide predicted, "Rovaniemi will change drastically when the Chinese tourists return … but now there are English, Italian, French, Israelis, Brazilians, Swiss … it was not expected that so many other people would come. Japanese and Koreans could also be groups attracted to the cold."

Pointing to the shifts in what attracts visitors, one representative of Visit Rovaniemi, the local tourism bureau, explained, "It used to be the midnight sun why people were traveling here in the beginning then it changed in the 80s to winter tourism. Maybe summer tourism will change when the weather in southern European regions becomes too hot, but that we don't know." While Rovaniemi has been able to capitalize on interest in the aurora and winter tourism, if Chinese tourists were to grow fonder, for instance, of Arctic cruises, this could divert flows of visitors from inland destinations to maritime spaces. Already, Norwegian cruise company Hurtigruten offers "China-friendly" sailings, with Chinese experts, translators, and chefs (Dragon Trail International 2021).

Yet another shift to tourism patterns in Rovaniemi could occur if the controversial Arctic Railway were built. The proposed corridor would extend the Finnish railway network from Rovaniemi to the Norwegian coast at Kirkenes, a port town which has welcomed numerous Chinese delegations over the past few years (Bochove 2024). In 2019, Chen Feng, General Director of COSCO's Marketing and Sales Department, remarked of the proposed Arctic Railway, "It would be welcomed by shipping companies." It would also likely be welcomed by Chinese tourists and the operators serving them in northern Fennoscandia. The project's boosters tout the possibilities of expanding tourism into new northern areas, but many are vehemently opposed, including the Sámi, whose customs are already heavily appropriated, commodified, and packaged for sale to tourists. In early 2021, the Lapland Regional Council rejected building the Arctic Railway until 2040 (Kähkönen and Nystén-Haarala 2021). Later that year, however, the council proposed a plan that would circumvent Sámi lands by connecting the municipalities of Kittilä and Sodankylä to the existing railway network between Kemijärvi, Kolari, and Rovaniemi. In their contribution to the debate, Kähkönen and Nystén-Haarala (2021: 6) state that this new route, which the council claims would promote tourism, mining, and forestry, might be "a trick to get the existing railway further north, a step closer to the melting Arctic Ocean." A new geopolitical context may also be motivating renewed interest in the railway. The war in Ukraine and associated concerns over critical infrastructure being built too close to Russia have prompted Finland to strengthen its connections with its western border and Sweden, including via the development of railway links. At the same time, some in Finland think the proposed Arctic Railway, which would terminate at Kirkenes, just 12 kilometres from the Russian border, could offer an all-too-easy inroads to an invading force.

Anticipating future Chinese tourism worldings in Finnish Lapland requires attention to both changing Chinese imaginaries of the Arctic and shifting infrastructural realities. Somewhere between the imaginary and the concrete—between dreams of auroral winter wonderlands and seamless intermodal connections—the Arctic world is remade in the vision of China, even as its origins remain hidden or unknown to those in its presence, such as the Indian visitors who now make use of infrastructure originally designed to support Chinese tourism. This points to how even if Chinese tourists never return in significant numbers, the world they created in Rovaniemi is already being absorbed into that of other visitors. Although "authenticity" is a central element of Arctic tourism, the industry continually reinvents itself, the experiences it offers, and the infrastructure used to do so. In the absence of Chinese tourists, Rovaniemi's tourism sector has adjusted its programs. One interlocutor noted that tourists from Israel have recently discovered Finnish Lapland as a safe place to travel. In Rovaniemi, to adjust to their kosher diets, special kitchens have been built and, reportedly, a Jewish rabbi flown into the city to offer his blessings. Many of these new groups of visitors, in addition to existing tourist groups, will likely make use of infrastructure such as glass igloos. In this way, Asian or Chinese tourism worldings are folded into other tourism worldings. This melding blurs the distinction between different Arctic imaginaries while gradually globalizing the Arctic and transforming it into a mirror of the rest of the planet.

Conclusion

In the years leading up to the pandemic, Chinese tourism represented a significant force that partially re-worlded Finnish Lapland. While the pandemic halted China's worlding capacities in the Arctic, their imprint is still visible in Rovaniemi. Tourism worlding, like any act of worlding, arises in the encounter between different groups of people, knowledge systems, and technologies. It exoticizes the destination while rendering it more familiar to visitors from afar. Such a transformation is risky when experiences of authenticity are so highly valued, yet it is inevitable, too. When tourists travel, they do not visit a place in a vacuum. Instead, their imaginaries and infrastructures influence the destination. The destination also continually and proactively reshapes itself, whether by offering package tours or packaged landscapes, all of which form a type of anticipatory infrastructure.

At Santa Claus Village in late June 2024, Christmas carols were chiming from the loudspeakers. A few tourists, posed on the Arctic Circle painted onto the ground, took photographs as proof that they had crossed the iconic line of latitude (66°30' N) demarcating the "real"

Arctic. Yet, with China considering itself a "near-Arctic state" in its 2018 Arctic Policy, one has to wonder where the Arctic—and, indeed, the Polar Silk world—begins and ends within Chinese imaginaries. Does the Arctic begin at the Arctic Circle, or does it begin somewhere closer to or even within China? Do Chinese "ice imaginaries" (cf. Lundén et al. 2023) include the Third Pole in the Himalayas, which constitute the world's third-largest storehouse of ice after the North and South Poles? How might Chinese tourists' experiences in the Himalayas shape their expectations for what they might encounter in the Arctic, and how might such experiences in the north "travel back" to places such as the Tibetan Plateau? What are the contributions of interactions between residents and tourism operators?

Research on the cultural and physical reshaping of the icy worlds visited by Chinese tourists, particularly by researchers with Chinese language skills and deep cultural familiarity, would broaden understanding of the Arctic's impacts on the rest of the planet across sociocultural dimensions. Such work would be especially compelling given the rise of the ice humanities (Dodds and Sörlin 2022) and the cold humanities (Kowal and Radin 2024). While it is widely understood that what happens in the Arctic is affecting the rest of the planet ecologically, there is less recognition of its capacity to influence the rest of the world socioculturally, too.

Ria-Maria Adams is a Postdoctoral Researcher in Social and Cultural Anthropology at the University of Vienna and a Visiting Researcher at the Arctic Centre in Rovaniemi. Her work explores Arctic transport infrastructures, youth well-being, and local initiatives in shrinking towns in the Fennoscandian Arctic, where she has conducted fieldwork since 2018.

Mia M. Bennett is Associate Professor of Geography at the University of Washington. She critiques the geopolitics of infrastructure development in the Arctic and orbital space and has run a blog on the Arctic, *Cryopolitics*, since 2009.

Acknowledgements

Our gratitude goes to the local people, tourism entrepreneurs and guides who shared their insights and knowledge with us. This publication was supported by the European Research Council project "InfraNorth" (PROJECT-ID: 885646).

Note

1. In contrast in the Asian Arctic, Chinese and East Asian encounters with Siberia, the Bering Strait, North Pacific, and wider Asian Arctic merit further critique and examination (e.g., Bockstoce 2009; Zatsepine 2017) to bring to life other "Arctic worlds" beyond the European one.

References

Adams, Ria-Maria. 2023. "Hunting a 'Good Life': Young Lifestyle Migrants in Finnish Lapland." *Polar Geography* 46(2–3): 139–56.

Augé, Marc. 1995. *Non-places: Introduction to an Anthropology of Supermodernity*. New York: Verso.

Bennett, Mia. M. 2015. "How China Sees the Arctic: Reading between Extraregional and Intraregional Narratives." *Geopolitics* 20(3): 645–668.

Bennett, Mia. M. 2015. "The Aurora I Didn't See." *Cryopolitics*. Retrieved August 1, 2025 from https://www.cryopolitics.com/2024/05/17/aurora-i-didnt-see/.

Bennett, Mia. M., and Benjamin L. Iaquinto. 2021. "The Geopolitics of China's Arctic Tourism Resources." *Territory, Politics, Governance* 11(7): 1281–1302.

Bernard, H. Russell. 2006. *Research Methods in Anthropology: Qualitative and Quantitative Approaches*. 4th ed. Lanham: Altamira Press.

Bhutia, Peden Doma. 2022. "Chinese Outbound Travel Will Look Like This When It Eventually Returns." *Skift*. Retrieved September 25, 2022 from https://skift.com/2022/05/19/chinese-outbound-travel-will-look-like-this-when-it-eventually-returns/.

Bochove, Danielle. 2024. "China's Arctic Dreams Make a Tiny Port a Global Prize." *Bloomberg*. September 6, 2024. Retrieved February 22, 2025 from https://www.bloomberg.com/features/2024-china-norway-russia-arctic-trade/.

Bockstoce, John R. 2009. *Furs and Frontiers in the Far North: The Contest Among Native and Foreign Nations for the Bering Strait Fur Trade*. New Haven: Yale University Press.

Cai, Yuan, Ma Jianyu, and Young-Sook Lee. 2020. "How Do Chinese Travelers Experience the Arctic? Insights from a Hedonic and Eudaimonic perspective." *Scandinavian Journal of Hospitality and Tourism* 2(2): 144–65.

Carse, Ashley, and David Kneas. 2019. "Unbuilt and Unfinished: The Temporalities of Infrastructure." *Environment and Society* 10(1): 9–28.

Cassinger, Cecilia, and Åsa Thelander. 2020. "Co-Creation Constrained: Exploring Gazes of the Destination on Instagram." In *The Routledge Companion to Media and Tourism*, ed. Maria Månsson, Annae Buchmann, Cecilia Cassinger, and Lena Eskilsson, 165–175. London: Routledge.

Chimirri, Daniela, and Carina Ren. 2022. "Tourism Worlding: Collective Becoming in East Greenland." *Polar Record* 58: e33.

CNN. 2024. "Overtourism Has Hit Santa's Village, Say Finnish Locals." CNN, December 13, 2024. Retrieved February 21, 2025 from https://www.cnn.com/2024/12/13/travel/overtourism-santa-claus-village-finnish-lapland/index.html.

Cole, Dermot. 2018. "How a Baseless Myth About Japanese Tourists and the Northern Lights Was Born—And Still Persists." *Arctic Today*, March 30, 2018. Retrieved April 17, 2025 from https://www.arctictoday.com/baseless-myth-japanese-tourists-northern-lights-born-still-persists/.

Degiampietro, Anja. 2024. "Rovaniemi Airport Celebrates Opening of New Flight Routes." Retrieved February 22, 2025 from https://www.nordicmarketing.de/en/blog/rovaniemi-airport-celebrates-opening-of-new-flight-routes.

Descola, Philippe. 2014. "Modes of Being and Forms of Predication." *HAU: Journal of Ethnographic Theory* 4(1): 271–280.

Dodds, Klaus, and Mark Nuttall. 2016. *The Scramble for the Poles: The Geopolitics of the Arctic and Antarctic*. Cambridge: Polity Press.

Dodds, Klaus, and Sverker Sörlin, eds. 2022. *Ice Humanities: Living, Working, and Thinking in a Melting World.* Manchester: Manchester University Press.

Dragon Trail International. 2021. "Hurtigruten: Chinese Travelers and the 'New Luxury.'" August 2021. Retrieved February 22, 2025 from https://dragontrail.com/resources/blog/hurtigruten-chinese-travelers-interview.

Falk, Martin, and Markku Vieru. 2017. "International Tourism Demand to Finnish Lapland in the Early Winter Season." *Current Issues in Tourism* 22(11): 1312–1326.

Finnish Road Safety Council. 2025. "Driving Tips to Finland." Retrieved February 21, 2025 from https://www.liikenneturva.fi/en/road-safety/finland-by-car/.

Gupta, Akhil. 2018. "The Future in Ruins: Thoughts on the Temporality of Infrastructure." In *The Promise of Infrastructure*, ed. Nikhil Anand, Akhil Gupta, and Hannah Appel, 62–79. Durham: Duke University Press.

Hall, C. Michael. 2008. "Santa Claus, Place Branding and Competition." *Fennia—International Journal of Geography* 186(1): 59–67.

Hall, C. Michael. 2014. "Will Climate Change Kill Santa Claus? Climate Change and High-Latitude Christmas Place Branding." *Scandinavian Journal of Hospitality and Tourism* 14(1): 23–40.

Handler, Richard, and William Saxton. 1988. "Dyssimulation: Reflexivity, Narrative, and the Quest for Authenticity in 'Living History.'" *Cultural Anthropology* 3(3): 242–60.

Havnes, Heljar. 2020. "The Polar Silk Road and China's Role in Arctic Governance." *Journal of Infrastructure, Policy and Development* 4(1): 121–38.

Herva, Vesa-Pekka, Alix Varnajot, and Albina Pashkevich. 2020. "Bad Santa: Cultural Heritage, Mystification of the Arctic, and Tourism as an Extractive Industry." *The Polar Journal* 10(2): 375–96.

Hong, Nong. 2014. "Emerging Interests of Non-Arctic Countries in the Arctic: A Chinese Perspective." *The Polar Journal* 4(2): 271–86.

Howell, Signe, and Aud Talle. 2012. "Introduction." In *Returns to the Field: Multitemporal Research and Contemporary Anthropology*, ed. Howell Signe and Aud Talle, 1–22. Bloomington: Indiana University Press.

Huang, Ming Feng, Tang Chuan Zhong, and David Weaver. 2017. "The Arctic Tourism Experience from an Evolving Chinese Perspective." In *Arctic Tourism Experiences: Production, Consumption and Sustainability*, ed. Young-Sook Lee, David B. Weaver, and Nina K. Prebensen, 89–99. Wallingford: CABI.

Iaquinto, Benjamim L., Mia M. Bennett, and Xiaofeng Liu. 2024. "The Biopolitics of Chinese Tourism Governance in the Arctic." *Geografiska Annaler: Series B, Human Geography* 106(3): 330–46.

Kähkönen, Juho, and Soili Nystén-Haarala. 2021. "Finland's Plans of the Arctic Ocean Rail Line are Buried Deep Beneath the Ice—or Are They Really?" Retrieved April 17, 2025 from https://lauda.ulapland.fi/handle/10024/64903.

Klinger, Judith, Mia M. Bennett, Ria-Maria Adams, and Eleanor S. Armstrong. 2023. "The Crowded and Empty Arctic: Examining Spatial Imaginaries and Infrastructures in Kiruna, Sweden. In *Arctic Dreams*, ed. Priebe Janina and Nina Wormbs. *Lychnos, Annual of the Swedish History of Science Society*: 57–72.

Kowal, Emma, and Joanna Radin. 2024. "Foreword: Cryopolitics after Ice" In *After Ice: Cold Humanities for a Warming Plane*, ed. Ruiz Rafico, Paula Schönach, and Rob Shields. Vancouver: UBC Press.

Lee, Young-Sook. 2020. "Asia and Arctic Tourism." *Scandinavian Journal of Hospitality and Tourism* 20(2): 105–9.

Liu, Nengye. 2018. "Will China Build a Green Belt and Road in the Arctic?" *Review of European, Comparative & International Environmental Law* 27(1): 55–62.

Lundén, Aapo, Alix Varnajot, Outi Kulusjärvi, and Mari Partanen. 2023. "Globalised Imaginaries, Arctification and Resistance in Arctic Tourism–an Arctification Perspective on Tourism Actors' Views on Seasonality and Growth in Ylläs Tourism Destination." *The Polar Journal* 13(2): 312–35.

MacCannell, Dean. 1973. "Staged Authenticity: Arrangements of Social Space in Tourist Settings." *American Journal of Sociology* 79(3): 589–603.

Martin, Natasha, and Carina Ren. 2020. "Adventurous Arctic Encounters? Exploring Chinese Adventure Tourism." *Scandinavian Journal of Hospitality and Tourism* 20(2): 126–43.

Mazzullo, Nuccio. 2022. "Issues of Sámi Representation in Finnish Tourism: A Quest for Authenticity." In *The Sámi World*, ed. Sanna Valkonen, Áile Aikio, Saara Alakorva, and Sigga-Marja Magga, 197–214. Abingdon and Oxon: Routledge.

Nyseth, Torill. 2009. "Place Reinvention at the Northern Rim." In *Place Reinvention: Northern Perspectives*, ed. Torill Nyseth and Arvid Viken, 1–14. Farnham: Ashgate Publishing.

Ong, Aihwa. 2011. "Introduction: Worlding Cities, or the Art of Being Global." In *Worlding Cities: Asian Experiments and the Art of Being Global*, ed. Ananya Roy and Aihwa Ong, 1–28. Malden: Wiley-Blackwell.

Saari, Ritva, Emily Höckert, Monika Lüthje, Outi Kugapi, and Nuccio Mazzullo. 2020. "Cultural Sensitivity in Sámi Tourism: A Systemic Literature Review in the Finnish Context." *Finnish Journal of Tourism Research* 16(1): 93–110.

Sinanan, Jolynna, Ria-Maria Adams, and Philipp Budka. 2025. "Framing Multipolar Tourism: Imaginaries, Visualities and Futures." *Visual Anthropology*, 38(1-2): 20–48.

Sokolíčková, Zdenka. 2022. "The Chinese Riddle. Tourism, China, and Svalbard." In *Asian Mobilities: Consumption in a Changing Arctic*, ed. Young-Sook Lee, 141–54. Abingdon and Oxon: Routledge.

Spivak, Gayatri Chakravorty (1985). "The Rani of Sirmur: An Essay in Reading the Archives." *History and Theory* 24(3): 247–72.

Stammler Florian, Aytalina Ivanova, and Vitebsky Piers. (this volume). "Vast Spaces and Inconsistent Anticipations in the Northeast Siberian Arctic." Oxford and New York: Berghahn Books.

Sukhorukov, V. D., Y. N. Gladkiy, A. A. Grigor'ev, and A. N. Paranina. 2021. "History and Geography of the Russian Arctic in the Domestic Travelogue." *IOP Conference Series: Earth and Environmental Science* 625(1): 012010.

Talvitie, Marjukka. 2022. "Rovaniemeltä Lennetään Eurooppaan Ensi Talvena Laajemmin Kuin Koskaan Aiemmin—Kohteina Esimerkiksi Pariisi ja Milano Sekä Uusimpana Bristol. [Next Winter Rovaniemi Will Fly to Europe More Widely Than Ever Before—Destinations Include Paris and Milan, and Most Recently Bristol." *Yle Uutiset*, July 18, 2022. Accessed April 17, 2025. https://yle.fi/a/3-12540723.

Tillman, Henry, Jian Yang, and Egill T. Nielsson. 2018. "The Polar Silk Road: China's New Frontier of International Cooperation." *China Quarterly of International Strategic Studies* 4(3): 345–62.

Wee, Desmond, Anthony Amato, and Marina A. Schmitz. 2021. "Where Do the Actors Live? The Chinese Tourist Gazes in Hallstatt." *Journal of China Tourism Research* 17(4): 570–91.

Zatsepine, Victor. 2017. *Beyond the Amur: Frontier Encounters Between China and Russia, 1850–1930*. Vancouver: UBC Press.

CHAPTER 5

Anticipating Arctic Roads in Sápmi

From Finnish Nation Building to Global Geopolitics

Veli-Pekka Lehtola

The region of Sápmi—the transborder Sámi area—lying in the northernmost municipalities of Finland, became integrated into the national road network from the 1910s to the 1950s. During the same period, Sámi culture experienced enormous changes. The transformations have often been described as the modernization of Lapland, Fennicization of Sámi culture, or Finnish colonialism of northern territories. The initially slow developments accelerated after World War II from the reconstruction period to "the snowmobile revolution" in the 1960s (Pelto 1973).

Building a road network to northern Lapland, and later to the Arctic Ocean, were projects initiated by the Finnish government, with their own backgrounds and purposes as well as consequences for employment in the region. The roads were tangible slashes in the landscape but also symbolized a new phase in the development of Sámi-Finnish "borderlands." New infrastructures were extended to the north in order to replace Sámi structures of land use, cultural traditions, and languages.

The Sámi people had traffic routes and mutual networks for centuries, with their own infrastructures and rationalities. In Sámi cartography, long-distance traffic was directed more to the north than to the south, since the coast of the Arctic Ocean was the primary goal for trade and influence. Changing these infrastructures was arguably a conscious objective for the Finnish state. Roads connected the Sámi territory more closely to the Finnish nation and started to integrate local cultures to Finnish society. They brought new people to the area—starting from road builders, then surveyors, mappers, administrators, and teachers—all Finnish. As old community structures were transformed, the Sámi had to find new ways to function in relation to Finnish society.

In this chapter, I examine histories of road in the first half of the 20th century in Finland as they come to bear on Sámi agencies and state visions of infrastructure today, in particular transport corridors to the Arctic Ocean. Northern Lapland and the Indigenous Sámi population was intensively integrated into Finland and to the Finnish nation by means of mapping, administrative efforts, and road building. This included a road to the Arctic Ocean during Finland's colonial acquisition of the Petsamo region. At the end of the 2010s, colonial and extractive associations with top-down road building in Sápmi during the 20th century resonated in protests against plans for an Arctic Ocean Corridor. Discussions about an Arctic Ocean route are ongoing with Finland's entrance into NATO, this time along an alternative path that follows the border areas between Sweden and Finland rather than passing through the most politically contested Sámi reindeer herding lands.

I argue that histories of top-down infrastructural building that usurped local decision-making, steer Indigenous agencies toward the resistance of future large-scale projects. This reveals the power of infrastructural memory to shape the entanglement of megaprojects with local agencies.

Roads and Ideology

Since Max Gluckman's study from Zululand, launching from the opening of a road bridge to describe complex ethnic relations, there has been growing ethnographic interest in the space "where diverse social and cultural groups move, meet, and interact" (Dalakoglou and Harvey 2012). Trails, paths, and roads are manifestations of human movement through the landscape and central to an understanding of mobilities on multiple scales (Snead, Erickson, and Darling 2010; Creswell 2011; see also Brody 1988; Huggan 1991).

Modern mobilities often shed light on the structures and operational modes of communities and societies, which otherwise would be considered stationary and constant, maybe even stagnant in a "traditional" phase. This connection has been emphasized in literature by both Indigenous and non-Indigenous scholars. Maori researcher Linda Tuhiwai Smith (2008: 52), for instance, points out that the building of roads, "the clearing of bush" (in our context, reorganizing land management in the Sámi area), and the establishment of missionary or trading stations were practices that involved the marking, defining, and controlling of space.

When contemplating the historical context for road building in Utah counties, Rogers (2013: 6–7) notes: "For some, the value of a road is its utility to move people and goods across the landscape or as an assurance of future economic development and productivity. Some base value on ease of transportation, commercial opportunity, or an ideology hinging on free and universal access or local sovereignty and tradition." Rogers himself, however, considers roads to be larger expressions of ideology rooted in history and culture:

> Beyond the physical, roads are cultural sponges—artifacts, timepieces of earlier times. If not in physical form, in cultural memory they reveal much about how generations, then and now, thought about the land and acted on it. The routes may have been created randomly, designless and haphazard, but each was premeditated. Before becoming a physical presence on the land, roads existed in imagination, and only later evolved as the products of long-fermented but contentious dreams. (Rogers 2013: 6–7)

According to Rogers, the road is "both material artifact and narrative device, both physical object and metaphor for confrontation or conflict." Roads in Utah county are "a winding, uncertain path into the heart of American culture's relationship to land and nature." Moreover, Rogers emphasizes that roads should be understood in the larger American "narrative of progress" (Rogers 2013: 6–7).

Roads clearly reflect the values of majority society. They were important for population growth, which was one of the main standards of the Finnish authorities for measuring the viability of the region. As Shirley Sponholtz (2012) has noticed, there has been a strong tendency already from the earliest times for "Western" economists to consider the existence—or nonexistence—of a road system and traffic connections to be one of the strongest indicators of a society's level of development. Besides the work of colonization and settlement, roads were essential for transportation or exploiting the region and its natural resources. In this way, it is possible to see roads as representing the embodied ideologies of progress and development in the landscape.[1]

In the development of networked infrastructures by the Finnish government, involving the building of roads, the establishment of telephone lines, and area mapping, the aim was to replace former infrastructures with new connections. The construction aimed to embody new standards and conventions of practice, for the purpose of administration and services in the first instance, and then increasingly for everyday life, including mobility and other social relations. New infrastructures reflected and drove historical concentrations of power and control, and they were instruments to reinforce institutional arrangements that changed former structures of relations and belonging (Dourish-Bell 2007: 416–17; Star 1999: 380–82; Masqulier 2002; Moran 2010).

In this way, road building was not just a technological innovation affecting mobility and economic circumstances in new ways. The phenomenon also reflected strong cultural changes because it impacted a whole network of local infrastructures, which had earlier shaped the ways of traffic and transport, trade, and social interaction within and between local communities, in other words, belonging—how to build "our" networks and interaction, and who will be "us" and who "them."

Demarcations, mapping, as well as road building reflected and constructed the values, hopes, and future visions that decision-makers had for Lapland and the Sámi territory—and consequently for the Sámi—although they were not expressed in words. Doreen Massey has noted that national maps, for example, do not reflect reality as such; they visually emphasize mutual agreement between nations and decision-makers' hopes in order to produce national clarity by means of borders.

Finnish maps from the early 20th century, where the Finnish border towards Norway was particularly highlighted, depicted the hopes of administrators for the unity of the Finnish nation, although this was far from realistic from the Sámi point of view (cf. Rautio Helander 2008). A mental border was being built between "us (Finns)" and "them (Norwegians)" by discarding the local reality of "these (Sámi)." Similarly, the road contained plenty of associations of hopes and values, although they were not easy to perceive at first glance.

In the Sámi area, roads were bringing new Finnish residents as settlers. In addition to progress on economic and administrative levels, the roads also had strong cultural effects. According to Marjut

Aikio (cited by Korhonen 1988), roads improved the possibilities of many Finnish institutions, such as postal services, health care, police, and border control—and "they all spoke Finnish." In other words, the language of the road was Finnish. However, the Sámi trails and routes, cultural networks, and structures, were still working under the surface of the new infrastructures. These parallel modes of land use and customary law brought continuous tension to Finnish–Sámi encounters, sometimes leading to contradictions and even collisions, debates, and conflicts, which interestingly reflected and exposed cultural boundaries. Histories of road then reveal multiscale tendencies and structures, which are not just sudden outbursts, but have developed over a long period of time. The ways these histories are remembered shape Sámi action in relation to new infrastructural proposals.

The Arctic Ocean Corridor

Associations of roads with top-down state dreams of progress, enacted through settlement and extraction, informed Sámi resistance to recent plans for a transport rail corridor to the Arctic Ocean. The Jäämeren rata, or "Arctic Ocean Corridor," was a resurrection of Finland's dreams of a route to the Arctic sea. Protests began in 2018, when the route materialized in preliminary maps for regional land use projected for the year 2040, proposed by Lapin Liitto—the Regional Council of Lapland. The plans featured a rail link from the "capital" of Lapland in Rovaniemi (see Chapter 4), to the port city of Kirkenes on the Norwegian coast, with transport connections to the south and potential trade links across the melting Arctic to Asian markets. While the Finnish and Norwegian transport ministries assessed viability, private entrepreneurs and investors lobbied for the infrastructure and its promises, seeking Nordic and Chinese investment for the project if government plans did not come to fruition (which later happened when the transport ministries' report declared the project economically nonviable). Chinese interest in the project stemmed from its potential as an infrastructural link of a "Polar Silk Road" (see Introduction).

The perception that state and global players were governing Sámi land use through planning bureaucracies without sufficient consultation, generated collective outrage in Sápmi.[2] A coalition of Sámi activists, reindeer herders, and Greenpeace Finland initiated protests that spread across Finnish Sápmi in 2018, with debates reverberating across the Norwegian border. In these counter visions to state and corporate plans, the railway was not only a means of transport—it foreshadowed the future proliferation of logging and mining that could further degrade lichen in reindeer pastures, along with the rail line that threatened to bifurcate grazing areas. Repeatedly, the past was evoked as a harbinger for the future; Sámi political leaders stated outright that one need only look to the impacts of past transport infrastructures to understand that this was just another top-down state project meant to degrade Sámi lands and resources to fuel wealth elsewhere.

Considering this turmoil of visions, Rogers' quote cited earlier is especially relevant: "Before becoming a physical presence on the land, roads existed in imagination, and only later evolved as the products of long-fermented but contentious dreams." In the case of the Arctic Ocean Corridor in Sápmi, the imagination of past top-down infrastructures fueled Sámi fears of future transport routes imposed by state, private, and foreign actors. This shaped the Sámi and environmental resistance movement that ultimately succeeded in (temporarily) halting plans. When discussions of a transport corridor to the Arctic Ocean were revived in the early 2020s with Finland's new NATO membership and Nordic security arrangements adjacent to a deepening iron curtain with Russia,[3] the route was diverted away from the core Sámi areas, along Finland's western border, through Sweden, and west to Narvik on the Norwegian coast.

The case reveals how infrastructural memory actualizes Indigenous agencies. When people remember infrastructure as the historical medium for state agendas of progress and development at the expense of local desires, emerging national and global transport dreams also become suspect of colonial, extractive purpose, thus mobilizing grassroots resistance.

Contested Mental Cartographies

A stereotypical mindset entails the mental aberration that the connection to the south had been a predetermined direction of influences. In reality, this direction seems to have been quite insignificant to the Sámi. We must remember that constructing roads from "Finland" to the Sámi territory—since the 20th century to today—has not only meant improving traffic connections but the creation of new networks that dismantle old ones by reshaping local cartographies. There have always been busy road networks and even main roads in the Sámi territory, but they were constituted and oriented in a different manner than later roads.

Nomadic reindeer husbandry was a way of life based on mobility. The basic unit, *siida*, was generally built on kin or family, who in autumn migrated *(johtit)* from the seashores to an interior forest area *(vuovdi)*. Spring started a new migration to the coast for the summer. Migration matched the rhythm of seasons and fulfilled the needs of the reindeer—to cool shores in summer, to the shelter of forest areas in winter winds. Migration was not random; siidas had regular pastures on the coast as well as in forest areas. Migration routes were also regular, although flexible in that they could be changed when necessary. The routes followed the terrain, and conditions were quite variable (Itkonen 1948 II: 228)

Over decades and centuries, both migration routes and pasture areas engendered many kinds of emotional bonds and places of remembrance, and the knowledge connected to them was conveyed to the next generations by parents and grandparents. With the annually recurring migration rhythm, it was characteristic of the remembering among Reindeer Sámi to organize experiences according to a spatial, not temporal logic.

Instead of historical chronology, memories kindled by places were more important to remembering—for example, what a known ancestor had experienced in that gully or at that campfire. Apart from place names, the tune and words or narratives of joiks (Sámi singing expressions) could bring back the meanings of places or important lessons or memories related to them. Oral tradition imbued the landscape and environment with memory. When poet Nils-Aslak Valkeapää (1994) later concretized the history of his own ancestors and the Sámi feeling of home, he wrote:

> Our ancestors kept fires on Allaorda
> on Stuorajeaggis' tufts
> on Viiddesčearru
> Grandfather drowned in the fjord while fishing
> Grandmother cut her shoe grass in Šelgesruohtu
> Father was born in Finjubákti in burning cold
>
> And still they ask
> where is your home[4]

In the Sámi "cognitive cartography," northern traffic networks and other connections were directed more to the north and the Norwegian coast than to the south. The "South" did not mean much on an emotional level either. While today's geopolitically formed plans of a transport corridor to the Arctic Ocean have destinations on the Norwegian coast similar to Sámi migration routes, the physicality, meanings, and practices of those routes are entirely different; roads and railways do not have the same places or practices of remembrance. In contrast, they threaten to fragment landscapes that not only sustain Sámi livelihoods but also convey cultural memory.

When Finnish cartographers lived for long periods in the Sámi territory at the beginning of the 20th century to carry out their work, they had time to discuss with locals and gain insight into Indigenous cartographies. Around campfires, the locals told them about reindeer husbandry and everyday life, but sometimes they also got carried away and recounted wild tales from the Finnmark coast and Arctic Ocean fishing, which many had experienced. The surveyors, in turn, talked about their conditions in the south and events in Helsinki. Finnish news read from newspapers hardly interested the campfire comrades, however: "Finland is another country for them, and Lapland another. Their homeland does not extend far to the south of the Saariselkä watershed" (Castrén 1908: 61).

Even towards the end of the 1930s, Finland was still an abstract concept in the Utsjoki area (located in the northernmost municipality of Finland on the border with Norway). When Finnish researcher Ilmari Hustich headed from Utsjoki over a fjell to the south and towards Inari (just 125 kilometers away, considered to be central Sápmi), he was told: "Greetings to Finland!" Hustich ironically called Utsjoki a "kingdom of its own" (Hustich 1946: 81).

Finnish civil servants wanted this situation changed, and that challenged the map image and traffic system of the Sámi territory. Mapping the Sámi territory and building roads aimed to constrict the empty area that still separated northernmost Lapland from the rest of Finland at the end of the 19th century. Nationalistic thinking emphasized the importance of building a united Finland, where also the remotest peripheries were integral parts of Finland's national property—not uncertain borderlands that Norway or Russia were ready to acquire into their spheres of interest.

Memories of Finnish infrastructural consolidation and the disruption of Sámi landscapes that ensued, surface with the periodic resurrection of plans for an Arctic Ocean Corridor. In the protests that followed the announcement of the railway, denouncements focused on the potential for the degradation and fragmentation of Sámi places, reindeer pastures, and migration routes. The latest Arctic Ocean Corridor plans with Finland's newfound NATO membership and fears of a Russian invasion with the war in Ukraine, likewise evoke memories of state security infrastructures treading over Sámi landscapes of remembrance. Military exercises that accompany the increasing securitization of the region, are coming into conflict with reindeer herding activities. This time it is not a consolidation of Finnish borders, but those of a united Nordic region, that are contesting local cartographies.

Histories of Arctic Ocean Roads

From the beginning of the 20th century, roads built by the state proceeded to the Sámi area, fueling strong transitions for Sámi societies in the Soabbat area (1910s), Inari municipality (1920s and 1930s), the Utsjoki region after World War II, and in Skolt Sámi territories in the Pechenga area (especially in the 1930s). Roads to the Arctic Ocean were particularly central to the Finnish imagination of progress and national consolidation. The realization of this infrastructure reordered Sámi life and agencies as people and industry streamed into the region. Although the route to the Arctic Ocean was ultimately lost with the ceding of Finland's colonial territory, its impacts continue to shape Sámi sentiment and action as plans for a transport corridor are revived today.

There was a remarkable road proposal in 1905 by the Lapland committee that aimed to enhance the economy of Lapland in the direction of the Arctic Ocean, while also strengthening connections to the south. Roads could tie northern Lapland to the rest of Finland. Exploiting Lapland's natural resources, especially forests, was an even more important goal.[5] Matti Enbuske (2010: 7–8) has stated that, apart from these reasons, the purpose was also clearly to bring "Lapps into civilization," which was considered the Finnish mission at the time.

After the Tartu peace treaty and the annexation of Pechenga in 1920, until the ceding of this land with the Second World War, Finland realized its dreams of sea access via the Pechenga Road to the Arctic Ocean that made use of the new territory connecting Finland to the Arctic coast. Traffic on this road increased exponentially. Vuotso (between Inari and Sodankylä), for example, became a significant transit point for streams of tourists. The history of the road also started to influence Aanaar Sámi culture (around the area of Inari) already in the 1910s. Increasing interest in the forests of Inari led to speculating transaction, and consequently many Aanaar

Sámi lost their homes at a ruinous price. The change in the population ratio was related to the Pechenga Road, which brought a lot of new Finnish inhabitants (see Lehtola 2012: 72–81).

The Finnish population grew rapidly especially in Kyrö village and the Ivalo region, and the number of people living on municipal aid and relief work likewise grew. The situation began to manifest itself as a change in the internal focus of the municipality, when the Muddusjärvi village—with Inari village as its center—gradually lost its status as the municipal center. The strengthening of the Finnish Kyrö meant decreasing Sámi influence in municipal activities and decision-making bodies.

The road was built through the Aanaar Sámi area to Nellim and on to Pechenga. The construction and the activities around it provided Aanaar Sámi with work for wages and improved connections to both Ivalo and Pechenga. Forest work in southern and northeastern Inari increased from the 1920s, and Aanaar Sámi quickly adopted new influences, for better or worse. This resulted in discussions about the survival of Aanaar Sámi culture, in which stereotypical questions about the "authenticity" and "corruption" of Sámi culture were raised. The way that Aanaar Sámi commuted at the boundaries of many cultures, as well as their carefree adoption of new influences, confused Finns, who thought that *true Sámi* should represent a single ancient cultural form.

The question of Sámi culture emerged again with conflict surrounding Arctic Ocean Corridor proposals in 2018. Sámi politicians and activists emphasized threats to Sámi culture, and the art activist group Suohpanterror created art suggesting similar destruction, embodied in images of reindeer herds being run over by speeding trains. Although also implicating problematic notions of authenticity, these articulations became ways of Sámi resistance against state decision-making. With Sámi rights to maintain their own culture enshrined in Finnish law, pitting culture against the railway was one way of disrupting state visions of infrastructural expansion.

The "Devil" on Roads and Footpaths

The imagination and experience of roads in Sápmi have long generated fears, anxieties, and resistance to the changes and nefarious influences they could bring. These histories shape present-day trepidations and protest surrounding plans for "silk roads" and other Arctic Ocean Corridors, and the imagination of their anticipated impacts across Sápmi.

Interesting reactions toward road building were already present from the first constructions in Sápmi. Reindeer herders living along the route were worried about their pastures, but there were also suspicions and concerns about Finnish influence ("lantalaisvalta") becoming stronger with the road. Benefits of the road were counterbalanced by fears of disease and bad influences. Inhabitants of Kaukonen wanted to reject the road coming through the village and proposed a route farther away, as Samuli Paulaharju (1963: 11) noted: "Then all kinds of vagabonds, runaways and gypsies (tattaraiset) could not find their way into houses if the road passed through forests." Road engineers did not listen to the villagers, however.

The same argument was used by inhabitants of the Sámi Purnumukka village, when road construction progressed to Vuotso in 1912. Local oral histories and knowledge told that the road would only bring commotion and disrepute. Legend has it that the villagers convened and decided to divert the road far enough past the village. Uula Hetta reportedly enlisted in the road work just to find out about the plans of the Finns and to guide them "to the right direction." There is no detailed information about Hetta's tactics, but eventually the road passed Purnumukka at a distance of several kilometers (for more see Lehtola 2012: 75–78).

The artist Andreas Alariesto has related the Purnumukka villagers saying that all sorts of malefactors and runaways would come along the road. In the introduction to a painting, Alariesto commented: "Some kind of a path through the forest went through the Purnumukka village. When the road was then built, it was taken two and a half kilometers past Purnumukka. The villagers agreed among themselves not to pass along the same path, so that there will not be the slightest path to the village" (Korhonen 1988). Some families, meanwhile, moved away from the (future) roadside to live in more remote villages (Väyrynen 2010: 81–83).

Concerning the Vuotso area, Marjut Aikio (interviewed by Korhonen 1988) has referred to a more negative cultural influence of the road. She estimated that "the arrival of the road with all ensuing consequences was the most important impact to a change in new linguistic relations of the Sompio people—the language of the road was Finnish. Road builders from the south even taught Finnish to the Sámi of Sompio. The language of the church, schools, traders and authorities was Finnish." Aikio had considered the development strongly assimilating towards Finnishness and the Finnish language.

The "fear of road" became visible when the municipal council of Utsjoki declined "the benefits of development" in the middle of the

1920s. It wanted to prevent the construction of the Finnish road to Utsjoki and proposed a special Lapp area status for the municipality. Finnish decision-makers interpreted these actions as separatist politics, a wish to disengage from Finland. The road became a symbol that reflected the Utsjoki people's concern about the effects of Finnish society.

At the beginning of 1925, the municipal council of Utsjoki filed an appeal to Finnish MP A.A. Neitiniemi to start "measures to fully protect Lapp settlement in the Utsjoki parish." The goal was no less than to prevent outsiders from moving to Utsjoki. It called for a decree in the migration law that would not allow Finnish-born persons to move to the municipality of Utsjoki without permission from the municipal council, apart from those coming as official or service persons of the state. This would enable "the region of Utsjoki municipality to be preserved as a proper Lapp area".[6] The municipal council also made other proposals, which appeared in headlines in the *Rovaniemi* newspaper: "Amazing news from Utsjoki." The council decided to send an application to the governor of Oulu province requesting that no road or telephone line should be built to Utsjoki. It wanted to prohibit the establishment of a border guard detachment in the village. The municipal council was also opposed to building an elementary school in the central village of Utsjoki.[7]

Problems from excess population manifested in southern Inari during the 1920s and were reported in provincial newspapers, which Utsjoki people also read. Road construction brought new populations to the north, initially road builders, some of whom stayed in Inari municipality to wait for employment. New settlement arose along the roads as people were attracted to logging sites and the gold frenzy, for example. It was not possible for most newcomers to become farm owners, and consequently the unemployed excess population had to live on municipal relief and relief work.

Bad influences brought by the road were often mentioned as an argument of the Utsjoki people themselves. "When the subject is road construction, Utsjoki people always point out that it would bring 'bad Finns' with their bad habits. There is hardly ever any other explanation," deplored school counsellor Alfred Salmela, who made an inspection trip to Utsjoki in the summer of 1926.[8] On the other hand, tradesmen feared that Finland's road construction would take away their income.

The proposals of the Utsjoki municipal council in 1925 seemed to be contrary to all the ideals that civil servants and Finnish media considered the foundation of a developing society—economic structure based on agriculture, a growing population, developed traffic networks, modern education, and organized border control. A similar ideal was that Finns always wanted to decide the direction of development for the Sámi. It was acceptable that the Utsjoki people were possibly lured by Norway or dissatisfied with Finland's services, but it would be a frightening option if the Sámi would try to define and control things themselves.

In the end, the Finnish parliament dismissed the initiative without discussion. It was commented with the sarcastic views of the committees that nobody was moving to Utsjoki anyway, when everybody wanted to go to Pechenga.[9] The later municipal court changed its opinion and welcomed the Finnish infrastructure. Only a circulating story remained about a Sámi stating that an evil spirit and every kind of nuisance would come along the road. The story ended with the reply of a road supporter: "Can it not come here anyway" (Hämäläinen 1945: 222–23) or "the devil can get here on footpaths, too" (Lindroth 1970: 82–83).

These histories show that Sámi resistance against roads is not only a modern phenomenon, and that many of the same fears of roads present through the 20th century have been reproduced with recent discussions of an Arctic Ocean Corridor. In particular, anxieties surrounding foreign influence and associated cultural damage circulated on social and news media. Speculation surrounding potential Chinese investment exacerbated the situation, fueling everyday sentiments and political resistance against an Arctic Railway.

Memories of roads and the fears they brought of cultural loss and the "foreign," fueled visions of the most recent iteration of the Arctic Ocean Corridor as a conduit for top-down ideologies and outside influence. These perceptions were then reproduced into the future—as Sámi art, protest, and political debate—to stop infrastructural plans. The entanglement of past, present, and future imaginings demonstrates the ways in which collective memory shapes local agency amid the anticipation of infrastructural development. When that memory is of state power usurping Indigenous decision-making, it becomes more likely that fears override hopes, and that infrastructure is actively resisted.

Borders, Maps, and Roads

Benedict Anderson (2006: 164) cited the *census, maps, and museums* as institutions of power that helped colonizers in Southeast Asia shape conceptions and categories for power structures in new colonial areas. He considered them deeply political means to colonial annexation. As Mihály (2014: 42) states, such mediums are not only there to inform us about real data, but they also translate it for us, providing their own interpretations.

When discussing the colonial history of Finland in the first half of the 20th century, I would like to elaborate Anderson's idea with another list: *borders, maps, and roads*, all reflecting representations of Finnish "us" and understandings of progress.[10] These constructions were intended to eradicate earlier modes of land use and attach the Sámi areas to Finnish infrastructures. They helped create new models that reflected Finnish administrators' ideas and visions for the future.

From the middle of the 19th century, Nordic countries divided the Sámi territory between themselves. Border demarcations between Norway and Finland under Russian rule (1852), as well as between Finland and Sweden (1889), plunged Reindeer Sámi society especially into crisis as new dwelling areas had to be searched hundreds of kilometers away. The independence of Norway (1905) and Finland (1917) started a solidification of these borders, so that there were two fervently nationalistic states face to face also across the northern areas, and they were highly suspicious and unwilling to negotiate with each other (Aarseth 1989). This had strong impact on Sámi policy in Finland.

As national borders became defined and demarcated, local civil servants felt it important to strengthen them. This was illustrated from the 1890s onwards by cartographers and forest surveyors who, among other things, increased the detail level of the Sámi territory map. Apart from identifying natural riches, such as forest reserves, the cartographers compiled a visual image of northernmost Lapland as an integral part of Finland, while beyond the northern border on the Norwegian side there was hardly anything but fjells and a few place names (for more see Lehtola 2012: 58–81; Nyyssönen 2007).

The transnational Sámi way of life, which manifested in Utsjoki, for example, as continuous and effortless connections to the Norwegian side, was considered problematic to the goal of Finnish national integration. The aim of the government was therefore to strengthen the borders by impeding and preventing all types of dealings—even the movement of reindeer—across state lines. In hastily built schools, teachers from the south unrolled maps of Finland for Sámi children to marvel at and taught what kinds of rivers or birds there were in southern Finland. The activities of administrators on the Norwegian side were similar, but from Norwegian starting points. Along the Deatnu river forming the northern boundary between Norway and Finland adjacent to Utsjoki, children from the same families learned different national ideologies, some rooted in Oslo and others in Helsinki.

The Pechenga area abutting the Arctic Ocean was an equally important region in the project of national consolidation, and Finland defeated Norway in that race: Finland acquired it in the Tartu peace treaty in 1920. The Skolt Sámi in the area were to experience that Finns considered Pechenga their "colony"—one that they had not had before. Finnish authorities hoped that the opened route to the Arctic Sea would become a means to elevate trade and industry to an international level. The Pechenga period remained short, however, because Finland lost the area to the Soviet Union in the aftermath of World War II in 1944 (for more see Lehtola 2012: 254–79).

But the dreams never left, resurrected again and again with new national and global developments. Borders, maps, and roads continue to be significant in Finnish desires for progress, territorial consolidation, and international connection. However, the contours of these desires have transformed.

Instead of solidifying a Finnish national border, the Arctic Ocean Corridor is emphasized as transnational conduit and connector. Potential sea access in Norway serves Finnish security and economic interests. The mapping of the Corridor in Lapland's 2040 Land Use Plan inscribed these transborder visions and the Nordic cooperation that promised to facilitate them. With the flare-up of geopolitical tensions stimulated by the 2022 escalation of war in Ukraine and Finland's entrance into NATO, the cooperation has further morphed to form a unified Nordic border against Russia. Roads are a means of materializing these new geopolitical alignments.

Instead of the national integration of the Sámi regions, which drove 20th-century road building, the most recent infrastructural visions have aimed to use the Sámi regions as a thoroughfare while increasing their cross-border interconnectedness. Similarities between past and present strategies include the maintenance of a south–north connection, and the easing of mining, logging, and other industry along the route. However, instead of tying Sámi land to southern Finland, and forming a border against surrounding nations, the focus is on linking northern parts of the country to adjacent Nordic partners and western allies via transborder transport networks and sea routes north and west.

Through these transformations in national ideology inscribed in conceptions of borders, maps, and roads, top-down decision-making on transport corridors through Sápmi has remained, and therefore

Sámi actors have maintained a defensive stance against Finnish infrastructural visions. Protest art highlights fragmentation and disruption, whether as a train overrunning reindeer herds, or pastures torn down the middle by foreign visions of connection. The top-down nature of Finnish infrastructural planning past and present fuels perceptions of roads as externally imposed fragmentation, emphasized by Sámi actors visually and rhetorically in resisting the plans.

Roads of Indigenous Agency

Indigenous scholars and historians have increasingly emphasized the need to shift focus toward the study of Indigenous agency, away from disempowering narratives of cultural loss due to colonial subjugation and modernizing forces. Dakota historian Philip J. Deloria (2004) states that some Indigenous peoples—more than we have been led to believe—leapt quickly into "modernity" already in early historical phases; this was not because they adopted political and legal tools from whites or because of acculturation or assimilation, but because of their own will and interest. This notion of Indigenous agency in modernization problematizes the concept of "Fennicization," or the adaptation and assimilation of Sámi culture to Finnish culture. It has often been considered that the Sámi adopted "global" influences from the south, for example through Finnish networks.

In recent decades, there has been a turn rejecting purely state- and majority-related approaches in historical narrative, and instead emphasis on *borderlands*, where numerous economic and political interests meet and collide. Even then, scholars have been criticized for simplifying an "enormously complex historical phenomenon" by emphasizing colonial perspectives, while the actions and agencies of Indigenous peoples are marginalized or ignored.

Sámi histories in Finland have also mostly been studied in light of Finnish influences, which have been considered a point of departure for research. Beyond interpreting Sámi histories under asymmetric power relations, subjugation, and colonialism, I focus on how Sámi actors in the borderlands responded to the challenges of state infrastructural expansion, benefitted from new influences, and crossed many kinds of borders. Sámi continue to co-opt state infrastructures today to achieve social, political, and economic visions. In the case of the Arctic Ocean Corridor, it is imagined rather than physical infrastructural plans that are mobilized to ensure a vitality of Sámi lands and livelihoods.

During World War II, new kinds of traffic connections were built in the Sámi area by the Germans for their supply preparations and attack operations. On the Finnish side, the main roads between Kaamanen and Karigasniemi as well as Kaaresuvanto and Kilpisjärvi passed directly through important Sámi settlements, which until that point had had less outside influence due to their locations in roadless areas. The constructions foreshadowed later developments following the war, when the use of the roads by Finns became a major source of Finnish influence.

In the Lapland War between Germans and Finns (1944–45), the retreating Germans destroyed all of Finnish Lapland like they destroyed Finnmark in Norway, especially along the roads. The first massive change was the evacuation of the population—Sámi included—to more southern Finland. The Sámi population came into everyday contact with Finnish culture for the first time, and the situation of the evacuees became a prologue to the development that would take place in destroyed Sámiland in the postwar reconstruction era.

The destruction caused by the Germans was the trigger for a multiple transition period when the devastated areas were reconstructed in the years 1945 to 1952. Reconstruction meant the strong development of traffic connections, rebuilding houses, and a rise in living standards. There were two parallel processes in Sámi society and culture, partly contradictory. One was changes in "traditional" Sámi culture and an acculturation process that has been called "Finnicization," or even assimilation. The other process was that of ethnopolitical mobilization that gave rise to a new kind of participation of the Sámi people in majority society (for more see Lehtola 2019; Nyyssönen 2007: 120–55).

"Finnicization" involved economic, political, and cultural processes that integrated Sámiland to the Finnish nation more strongly than before. During the reconstruction period, new infrastructures, especially brand-new road systems, were built to connect the northern "peripheries" to Finland and loosen local connections to Norway. Again, roads made it easier to travel and transport materials; this was important, for example, when exploiting the natural resources of the north such as forests. The network of roads brought benefits such as an improved standard of living, and made Lapland more socioeconomically equal to other parts of Finland. However, the roads also had a reverse side: they decreased local "autonomy" as the Sámi territory was seen as a resource store. Its possibilities were largely viewed from the outside and often from an economic perspective (Lehtola 2012: 443–45).

The history of roads was reflected in the different speed of changes in different areas: the boundary between devastated and non-devastated areas after the war depended largely on traffic networks. Areas spared from destruction were distant from the roads that German patrols used when destroying settlements and dwellings. In these areas, life quickly settled into its previous course after the evacuation time. Martti Linkola (1990: 15) has stated that it was not until the economic and social changes of the 1950s that the Enontekiö wilderness houses and villages became similarly developed as the rest of Sámiland.

On the other hand, life often changed quite rapidly in the destroyed areas that had been accessible to the Germans. Newcomers settled along the roads that were constructed during the German presence; war veterans, mobile workers, and homeless wanderers came to work as "professional" carpenters and constructors. Many of them stayed in the north, and because all the newcomers were Finnish, the percentage of Sámi decreased radically.

The roads improved the operation of many Finnish institutions, such as postal services, health care, police, and border control, while making the Finnish language dominant in local administration and services (Korhonen 1988). Finnish centers, such as Ivalo, expanded almost explosively in the Sámi regions. The new center of Karigasniemi in Utsjoki municipality emerged by the roadside around the new border station near the old Sámi Gárigasnjárga. Due to border restrictions, the direction of trade turned from Norway to southern Finnish Lapland. Reconstruction itself symbolized—and partly instigated—the trend of Finnicization.[11]

Even the remote Deatnu region was rebuilt with Finnish standardized houses. Sámi residents had to march to the reconstruction office to look at standard drawings of a few available house types. The new houses were detached homes furnished according to the Finnish-Ostrobothnian style. Saunas with a dressing room and porch were also a novelty, although there had been a few even before the war. The Sámi tried to adapt to the new kind of houses by adopting Finnish material culture and practices. Utensils and decorations, curtains, and tablecloths started to appear in Sámi houses. Only the layout of the yard resembled the pre-war Lapland houses. In the words of Ilmari Hustich (1946), the loose placement of storehouses, shelters, and racks reflected "the subconscious tendency of the Sámi to keep something of the openness and space that prevailed around a hut camp."

Finnish ideals started to appear elsewhere in Sámi everyday life and customs. A change in clothing was one of the most apparent signs of new influences as Finnish clothes started to replace Sámi dress. Although this was partly due to the better availability of Finnish clothes during the post-war period with improved roads, it was also about fashion. Young people abandoned Sámi dress and dressed in "southerner clothes," especially when they were going to the central villages. The same was true with the language; speaking Sámi felt quite unfashionable and betrayed that the speaker was "a Lapp." Parents started to give their children Finnish first names and some also changed their surnames. Such Sámi attitudes and choices, fueled by Finnicizing forces, percolated into the administration of the Sámi region even without the introduction of Finnish workers into governmental offices; the meeting language of the Utsjoki municipal council, for example, changed to Finnish at the beginning of the 1950s, although almost all members were Sámi (Lehtola 2019: 170–220).

A strengthening ethnopolitical movement reflected the other side of Finnish infrastructural expansion for the Sámi, as the increased Finnish cultural and institutional presence brought by roads created an arena to adopt new strategies and resources. An obvious example of the twofold process is the question of education in relation to Sámi culture, reflecting well the relevance of expanding road networks. The school system has been considered a major Fennicization factor in Finnish Sápmi at the end of the 1940s. A new compulsory education act in 1947 obliged even the inhabitants of remote areas to send their children to year-round schools in central villages. This mandate was made possible by improved transport infrastructure, wherein lay a paradox: the roads were considered good enough to separate children from their homes but not good enough to bring them back except during longer holidays, such as Christmas, Easter, and summer holidays.

The children were gathered at large school centres with dormitories, which was in itself a shock to self-esteem. The trauma of many Sámi children resembled the negative evacuation experiences also facilitated by roads. Spending whole winters in a Finnish-speaking environment away from their home districts estranged the children from their cultural backgrounds. Learning and adopting the Finnish language and constantly cramming the ideas of Finnish culture from Finnish national literature in school, the children were guided to the Finnish language, culture, and worldview. The Sámi started to think in Finnish and as Finns. The children's natural connection to their home language was severed, and skills that had been conveyed by everyday practice—such as reindeer husbandry and crafts—were forgotten or never learned (Lehtola 1997: 62; Sara 1984: 41–43).

The other side of the coin was that education also brought new notions about the role of the Sámi as equal citizens of Finland and as members of international networks. It was the first educated "boarding house generation" that recognized that the Sámi issue involved larger circles than just the local family and kin-group area. To compensate for the lack of power at local and national levels, the Sámi movement orientated itself more and more to global arenas, especially to Indigenous discourses in the 1970s. They raised questions about their rights as a minority, ethnic group, and Indigenous people. The fight to substantiate their land usage "from time immemorial" started. Sámi actors adapted global trends in multiple ways, depending on their usefulness and applicability in local contexts, including relations to other ethnic groups (Nyyssönen 2007: 182–210).

In these ways, besides problematic changes, roads created new possibilities for the Sámi to influence majority society both as individuals and as ethnopolitical actors. Many studies have emphasized that early Indigenous responses to colonial infrastructural expansion were not necessarily those conventionally labelled as "resistance." As May (2015: 53–54) states, instead of seeking just "readily recognizable forms of resistance," it is important to notice the diverse forms of resistance. According to Lester and Laidlaw (2015: 11–12), it could be a more "tenacious adaptation that carries with it its own, more subtle form of resistance." They point to resilience and persistency especially on the micro level, inside families and small communities. Also in Finland, we can see that although the Sámi did not openly resist the new Finnish order, they were not such helpless "victims of development" as has often been suggested. Instead, we should understand the cultural response on a larger scale, going beyond the concept of resistance. Sámi societies have had multiple strategies to deal with outside influences and novelties.

Likewise in recent debates surrounding the Arctic Ocean Corridor, Sámi agency has been central not only in resistance against the railway but in shaping public perception, discourse, and memory surrounding infrastructure in Sápmi. Sámi activists and reindeer herders, together with environmentalist allies, were successful in halting and redirecting the proposed route away from core reindeer pastures by mobilizing memories of infrastructural encroachment in the region. As an Arctic Railway was proposed top-down via state and regional governing bodies, memories of past cultural and material damages resulting from infrastructure building were evoked to challenge the plans. Transport infrastructure in Sápmi was remembered as a means of colonial settlement, resource extraction, and cultural and linguistic assimilation. These memories became harbingers of the logging, mining, and harm to Sámi reindeer herding lifeways that could follow an Arctic Ocean Corridor. Although negative effects were evoked over double-sided impacts, the infrastructural blueprints nevertheless became Sámi tools for new political agencies in determining regional land use.

Interplays of past and present infrastructural development in Sápmi inform broader understandings of how memories of infrastructure shape local anticipations and agencies. Memories of roads as externally imposed, and new plans for large-scale transport corridors driven by national or foreign investment, feed each other, fueling Indigenous agency in determining infrastructures and actualizing their political potential.

Veli-Pekka Lehtola is Professor Emeritus of Sámi Culture in the Giellagas Institute at the University of Oulu, Finland. Lehtola is a (North) Sámi from Aanaar (Inari) in northern Finland. As a researcher, Lehtola is specialized in the history of the Sámi and Lapland, in modern Sámi art, and in Sámi representations.

Portions of this text have been reproduced from other work by Veli-Pekka Lehtola including "Second World War as a Trigger for Transcultural Changes among Sámi People in Finland" in *Acta Borealia*.

Notes

1. In his dissertation *The Car Society*, Per Lundin uses the concept of "frozen ideology" by the Swedish historian of science and ideas, Sven-Eric Liedman, to depict the mid-20th century building craze in Sweden and its technical artifacts such as parking lots, parking houses and huge shopping malls. Bennesved uses the same concept about the roads built by states to rural environments, saying that at the beginning they are sketched as an abstract idea in plans and maps, taking on physical existence later—thus, representing embodied ideology in the landscape. See Bennesved 2014: 45–46; Sörlin 1988.
2. See Magnani, Natalia, forthcoming. "Sámi Land or Great Northern Forest? Mapping Indigenous and Environmental Outrage in Sápmi." In *What on Earth! More-than-Human Ethnographies in the Anthropocene*, ed. Crawford, Peter and Nils Bubandt.
3. See Magnani et al., forthcoming. "Material Imaginaries of the New Iron Curtain." *Social Anthropology.*
4. Valkeapää 1994. There are lots of Indigenous studies referring to the priority of spatial rather than temporal thinking related to places, see e.g., Deloria 1973: 62–77, 122; Kidwell, Noley, and Tinker 2001: 44–48. Sámi

memory has not been studied much from this angle, although these observations are quite familiar for example in toponym research, which is central in Sámi studies. Reindeer nomadism in its widest form ended in the 19th century with national border closures, which severed centuries-old migration routes and forced reindeer herders living in different nations to remain within national borders, with the exception of the arrangements on the border between Sweden and Norway. The Reindeer Sámi in the Enontekiö area continued their nomadic way of life after the border closures, though they could no longer go to the Norwegian coast.

5. Komiteamietintö [Committee Report] 1905, 236–47. "Komitealta Lapinmaan taloudellisten olojen tutkimusta varten [From the Committee for the Study of the Economic Conditions of Lapland]." Helsinki: Valtioneuvosto.
6. A. A. Neitiniemi's proposal "Yhtenäisen suunnitelman valmistamisesta valtakunnan lapinkielisen väestön taloudellisen ja sivistyksellisen tilan kohottamista varten [On the Preparation of a Unified Plan for Raising the Economic and Cultural Status of the Lapp-Speaking Population of the Country]." Rovaniemi February 2 1925. From archives: February 18, 1925. Liitteet 1V, 2. Anom. ehd. no 4. A. A.. Vp 1925; Valtiopäiväasiakirjat.
7. Rovaniemi May 11 1926.
8. Rovaniemi February 24 1925.
9. Rovaniemi February 18 and February 26 1926.
10. In Lapland, the census already meant an old tradition of categorizing people, and museums were not yet topical for the colonial reality in the field, except as a research task for some lappologist scholars.
11. Komiteamietintö [Committee Report] 1952. no 12, 89: "Saamelaisasiain komitean mietintö [Committee Report on Sámi Issues]." Helsinki: Valtioneuvosto.

References

Aarseth, Bjørn. 1989. *Grenser i Sameland [Borders in Sámi Land]*. Oslo: Norsk Folkemuseum.

Anderson, Benedict. 2006 (1983). *Imagined Communities. Reflections on the Origin and Spread of Nationalism*. London and New York: Verso.

Bennesved, Peter. 2014. *The Road in the Park: State Power and Ideology during the 20th Century Seen Through Maps of the Swedish Subarctic Abisko*. Umeå: Umeå University.

Brody, Hugh. 1988 (1981). *Maps and Dreams*. Vancouver: Douglas & McIntyre.

Castrén, Lahja. 1908. "Inarijärven Saarten Maantieteellinen Kartoitus v. 1906 [Geographical Survey of the Islands of Lake Inari in 1906]." *Maanmittariyhdistyksen Aikakauskirja (Land Surveyors' Association)*.

Creswell, Tim. 2011. "Towards a Politics of Mobility." In *Routes, Roads, and Landscapes*, ed. Mari Hvattum, Brita Brenna, Beate Elvebakk, and Janike Kampevold Larsen, 163–77. Surrey: Ashgate.

Dalakoglou, Dimitris, and Penny Harvey. 2012. "Roads and Anthropology: Ethnographic Perspectives on Space, Time, and (Im)mobility." *Mobilities* 7(4): 459–65.

Deloria, Philip Jr. 2004. *Indians in Unexpected Places.* Kansas: University Press of Kansas.
Deloria, Vine Jr. 1973. *God Is Red: A Native View of Religion*. Lakewood: Fulcrum Publishing.
Dourish, Paul, and Genevieve Bell. 2007. "The Infrastructure of Experience and the Experience of Infrastructure: Meaning and Structure in Everyday Encounters with Space." *Environment and Planning B: Planning and Design* 34: 414–30.
Enbuske, Matti. 2010: "Suomen Suurimmilla Tietyömailla. Valtamaantien Rakentaminen Kohti Inaria [At Finland's Largest Road Construction Sites. Construction of a main road towards Inari]." *Raito* 1: 4–10.
Huggan, Graham, 1991. "Maps, Dreams, and the Presentation of Ethnographic Narrative: Hugh Brody's Maps and Dreams and Bruce Chatwins the Songlines." *Ariel: A Review of International English Literature* 22(1): 57–69.
Hustich, Ilmari. 1946. *Tuhottu ja Tulevaisuuden Lappi [Destroyed and Future Lapland*]. Helsinki: Kustannustalo.
Hämäläinen, Antti. 1945. *Tunturien Mailta [From the Fell Lands]*. Helsinki: WSOY.
Itkonen, Toivo Immanuel. 1948. *Suomen lappalaiset* [*The Lapps of Finland. II*]. Helsinki: WSOY.
Kidwell, Clara Sue, Homer Noley, and George Tinker. 2001. *A Native American Theology*. Maryknoll: Orbis Books.
Korhonen, Jorma. 1988. "Saame Vaihtuu Suomeen. Pohjoisen Maaton Kieli ei Enää Välity Sukupolvelta Toiselle, Vaan Kuolee [Sámi is Changing to Finnish. The Northern Language Is No Longer Passed Down from Generation to Generation, but is Dying Out]." *Helsingin Sanomat*, 3 March.
Lehtola, Veli-Pekka. 1997. *Saamelainen Ääni. Saamen Radio [The Sámi Voice. Sámi Radio] 1947–1997]*. Helsinki: Yleisradio.
———. 2012. *Saamelaiset Suomalaiset—kohtaamisia [Sámi, Finns—Encounters] 1896–1953*. Helsinki: SKS.
———. 2019. *Surviving the Upheaval of Arctic War: Evacuation and Return of the Sámi People in Sápmi and Finland During and After the Second World War.* Inari and Oulu: Puntsi Publisher.
Lester, Alan, and Zoë Laidlaw. 2015. "Indigenous Sites and Mobilities: Connected Struggles in the Long Nineteenth Century." In *Indigenous Communities and Settler Colonialism: Land Holding, Loss and Survival in an Interconnected World*, ed. Zoë Laidlaw and Alan Lester, 1–23. New York: Palgrave Macmillan.
Lindroth, Saimi. 1970. *Terveyssisarena tunturipitäjässä [Nurse in the Fell Owner's House]*. Mäntsälä: Omakustanne.
Linkola, Martti 1990. "Ihmisten ja Porojen Maa [The Land of People and Reindeer]." In *Lapin kuvat*, ed. Marja Vuorelainen, 13–31. Helsinki: SKS.
Masqulier, Adeline. 2002. "Road Mythographies: Space, Mobility, and the Historical Imagination in Postcolonial Niger." *American Ethnologist* 29(4): 829–56.
May, Vivian. 2015. *Pursuing Intersectionality, Unsettling Dominant Imageries*. New York: Routledge.
Mihály, Vilma-Irén. 2014. "The Census, Maps, and Museums, or a Way of 'Translating Identity.'" *Philologica* 6(1): 35–43.
Moran, Joe. 2010. "The Future of Roads." *Soundings* 44: 107–16.

Nyyssönen, Jukka. 2007. "'Everybody Recognized That We Were Not White'. Sami Identity Politics in Finland, 1945–1990." Ph.D. Dissertation. Tromsø: University of Tromsø.

Paulaharju, Samuli. 1963(1923). *Wanhaa Lappia ja Peräpohjaa [Old Lapland and Peräpohja]*. Helsinki: WSOY.

Pelto, Pertti J. 1973. *The Snowmobile Revolution: Technology and Social Change in the Arctic*. Long Grove: Waveland Press.

Rautio Helander, Kaisa. 2008. "Namat dan nammii. Sámi báikenamaid dáruiduhttin Várjjaga guovllus Norgga uniovdnaáiggi loahpas [Names for this Name. Norwegianization of Sámi Place Names in the Várjjaga Area at the End of the Norwegian Union Period]." Ph.D. Dissertation. Kautokeino: Sámi allaskuvla.

Rogers, Jedediah S. 2013. *Roads in the Wilderness: Conflict in Canyon Country*. Salt Lake City: University of Utah Press.

Sara, Iisko. 1984." Suomalaisempia kuin Suomalaiset Itse [More Finnish than the Finns Themselves]." In *Bálggis—Polku. Lapin Sivistysseura 1932–1982 [Bálggis—Path. Lapland Cultural Society 1932–1982]*. Helsinki: Lapin Sivistysseura.

Snead, James E., Clark L. Erickson, and Andrew J. Darling. 2010. *Landscapes of Movement. Trails, Paths, and Roads in Anthropological Perspective*. Philadelphia: Pennsylvania University Press.

Sponholtz, Shirley. 2012. "A Brief History of Road Building." Triple Nine Society. Retrieved February 15, 2025 from https://www.triplenine.org/Vidya/OtherArticles/ABriefHistoryofRoadBuilding.aspx.

Star, Susan Leigh. 1999. "The Ethnography of Infrastructure." *American Behavioral Scientist* 43: 377–391.

Sörlin, Sverker. 1988. "Järnvägen som kulturbärare [The Railway as Culture Bearer]." In. *Malmbanan 100 år [The Ore Line 100 Years]: 1888–1988*, ed. Kjell Lundholm, 11–92. Luleå: Norrbottens Hembygsförbund.

Smith, Linda Tuhiwai. 2008. *Decolonizing Methodologies: Research and Indigenous Peoples*. London: Zed Books.

Valkeapää, Nils-Aslak. 1994. *Trekways of the Wind*. Tucson: University of Arizona Press.

Väyrynen, Anna-Liisa. 2010. "'En Tahtonut ois Vaihtaa tätä Vieraan Maisemiin'—Allasevakkojen Elämänvaiheista ['I Wouldn't Trade This for a Foreign Landscape']." In *Kylä Kulttuurien Risteyksessä. Artikkelikokoelma Vuotson Saamelsista [A village at the Crossroads of Cultures. A Collection of Articles about the Vuotso Sámi]*, ed. Ulla Aikio-Puoskari and Päivi Magga, 80–91. Vuotso: Vuotso Sámiid Searvi.

CHAPTER 6

The Expansion of the Port of Prince Rupert

Between Economic Development and Alternative Life Paths

Giuseppe Amatulli

"How is life in Prince Rupert? Oh, life is good. Very good. You know, Prince Rupert is a nice place. What about the port? Is it expanding? How is it going? Oh yes, it is growing. It is going well."[1]

Located on Kaien Island, one of the northwestern isles of British Columbia, Prince Rupert is home to the deepest natural harbor on the west coast of North America. In addition, its strategic location offers the shortest shipping route from North America to Asia. The combination of these factors has played a major role in the development of the port in the last fifty years, with Prince Rupert's dock currently being the third largest in Canada by cargo tonnage and containers handled.[2] Whereas not an Arctic town, Prince Rupert's strategic location with connections to the Arctic and sub-Arctic, its proximity to Alaska, and long-term envisioning as a hub of global transport and exchange with Asia, evoke imaginaries of "Arctic Silk Roads" parallel to those discussed in other chapters in this volume. Such imaginaries have been strengthened following the construction of the Fairview Terminal, the first dedicated intermodal container terminal in North America, operated by DP World. New hopes about development and economic growth have started to circulate among politicians, business owners, and locals. Such hopes revamped in 2022, when

DP World announced it had entered a two-year agreement with the Prince Rupert Port Authority to conduct an assessment regarding the feasibility of building a second innovative container terminal.[3]

Economic actors reacted with growing excitement to the possibility of seeing Prince Rupert grow through the port expansion; however, not everyone agrees that the relentless port development is the only available path. Whereas large-scale infrastructure development plans have been Prince Rupert's defining element in recent decades, smaller and unexpected economies have emerged around the imagined development of Prince Rupert. Originating from First Nations knowledge of food, harvesting practices, and fishing methods; a new economy based on local products and zero waste is emerging in Prince Rupert, with an abundance of small shops that promote local products as a way to assert ideas of sustainability intertwined with cultural continuity. In particular, the production and consumption of local food initiated by the Urban Farm Project supported by Ecotrust Canada, may be regarded as a way to raise awareness and encourage the search for a new balance based on the sustainable use of what is already available locally, in opposition to large-scale development (such as the port expansion), perceived as unsustainable and responsible for the continuous extraction of local resources.

By drawing from my fieldwork experience, this chapter aims to shed light on how the expansion of the port is shaping life in Prince Rupert, the expectations it generates, the impact it has on how people envision the future (and the kind of future they want), and how economies of anticipation occurring among local resident can counteract large-scale infrastructure development.

A Historical Overview between Past and Future Development in Prince Rupert

The idea of having a hub that could connect North-western British Columbia to Asia traces back more than 150 years ago. Already before the creation of Canada as an independent country, discourses around the necessity to connect what is now known as British Columbia with Asian markets were present. Once the steam railway became a reliable and economically effective way to move goods, discourses surrounding the possibility of having a North American 'land bridge' from the Atlantic to the Pacific to link up with ships bound for Japan and China were revamped (MacKay 1986: 7). In 1851,

Allan Macdonell, a mining promoter and former cavalryman, stated: "… Having its Atlantic seaport at Halifax and its Pacific depot near Vancouver's Island, it would inevitably draw to it the commerce of Europe, Asia, and the United States, so thus British America, from a mere colonial dependency, would assume a controlling rank in the world" (*New York Tribune*, March 27, 1851).

Trade between British Columbia and 'the Orient' started in the late 1860s. The completion of the Canadian Pacific Railway, extended to Vancouver in 1887, strengthened trade opportunities, which, however, remained for a long time limited to timber that was the main good exported from Canada; while silk and tea were imported from China and Japan (MacKay 1986: 110). Once the Canadian Pacific Railway was completed (1885), Charles Hays, general manager of the Grand Trunk Pacific Railway (GTPR), started to explore the possibility of building a second transcontinental railway (Wilson and Bowles 2016: 120). The idea was to connect Winnipeg with a place that "would surpass Vancouver as the dominant port of the North Pacific Ocean" (Prince Rupert Port Interpretive Centre 2022). The natural harbor on Kaien Island was identified as the ideal railway terminus and the starting point of a new Silk Route to Asia (MacKay 1986: 8).

Prince Rupert was founded in 1906 with the idea of boosting such trade, eventually taking over the port of Vancouver. Among the very first projects on how to develop the port, there was the idea of receiving silk from China and running silk trains from Prince Rupert to New York, to deliver silk thirty-six hours faster compared to trains starting from Vancouver (MacKay 1986: 102). The trading vocation of the city is well expressed in its coat of arms, which reads: "By rail and ship, with net and pick, we win our wealth" (figure 6.1).

Nevertheless, the opening of the Panama Canal in 1914 gave a tremendous boost to the port of Vancouver, with lumber companies able to ship their products at a competitive price to the US and the UK. In such context, Prince Rupert and its port struggled to compete, with the city never experiencing the boom that experts predicted after the completion of the Grand Trunk Pacific Railway (MacKay 1986: 118).

The trade between Canada and Asian Pacific countries increased throughout the 1920s, and in 1929 it amounted to $87 million in exports and $33 million worth of imports; with Vancouver that was considered the leading Pacific port. In such a context, also the port of Prince Rupert experienced some growth, with the first Japanese-owned grain ship arriving on Kaien Island in 1926 (MacKay 1986, 139–40). Nevertheless, the Great Depression and the political developments of the 1930s completely changed the scene. The full-scale war Japan initiated against

Figure 6.1. *Coat of arms of the city of Prince Rupert painted on the ferry to the city, BC, Canada, 2022. © Giuseppe Amatulli.*

China in 1937, and the beginning of World War II accelerated such changes, and by the end of 1941 all trade with Japan shut down. More than twenty years of effort to promote Asian Pacific trade was then halted, with the port that was built for Asian trade, Prince Rupert, not benefitting at all from such trade (MacKay 1986: 151).

However, once the Japanese threat to invade North America became real, the US Army decided to use Prince Rupert, and its railway, to move supplies and soldiers to Alaska. In a matter of months, 75,000 soldiers arrived and, due to its proximity to Alaska, a highway to Prince Rupert was built, together with two forts and an airbase, with the shipyard repairing 400 ships during the war and building a dozen 10,000-ton merchant ships, four minesweepers, and the first CN armored train to patrol from Prince Rupert deep to the interior (MacKay 1986: 152–53). After World War II, there was an attempt to establish a new shipping route between Shanghai and Prince Rupert. However, with the Chinese Communist Party rapidly taking control over China, such options were soon abandoned. Prince Rupert was cut off again, "robbed of an important trade potential," as the new president of the CN Railways, Donald Gordon, stated in 1950 during his first trip to Prince Rupert (MacKay 1986: 157).

Whereas Prince Rupert was cut off, the Pacific coast was not. In 1951, Japan signed the peace treaty, which also allowed the country to conduct its own trade again. Food and raw materials were needed, and quickly Japanese imports from Canada increased to $100 million per year. A few years later, in 1954, Japan and Canada signed a commercial agreement, with Japan becoming the third commercial partner of Canada, after the US and the UK (MacKay 1986: 158). Ten years later, in 1965, Japan exports overcame its imports for the first time since World War II, and a joint Canada–Japan governmental committee gathered to discuss Canadian raw material exports to Japan. By then, the volume of cargo through the port of Vancouver tripled to 30 million tons per year, and by the beginning of the 1970s, Vancouver was one of the busiest ports in the world. It was then clear that another port in British Columbia would need to be identified to serve the expanding trade between Canada and Asia. In such context, the idea of using the Port of Prince Rupert to support the trade with Asia started to be (re)consider (MacKay 1986: 170–73).

In such a socioeconomic context, Prince Rupert's citizens did not forget the Hays' vision of a 50,000-inhabitant city that could serve as a gateway to Asia. In 1964, a Port Development Commission was established, with the aim to foster the development of the port while lobbying at the provincial and federal levels (MacKay 1986: 177).

The Commission received support from nearby provinces, with the premiers of Alberta and Saskatchewan, Manning, and Tatcher advocating for a second west coast port. In particular, Manning recalled one of his trips to Japan, stating that key industrial figures were interested in the development of the Port of Prince Rupert for three main reasons: the increasing traffic in Vancouver port, the ability of the dock of Prince Rupert to receive large vessels, the savings in sailing time between Japan and Prince Rupert (MacKay 1986: 179). Despite these lobbying efforts, the Commission did not receive the support of any of the federal transport ministers who served in those years, who were not convinced that Prince Rupert could have enough traffic to justify relevant investments by the federal government. However, when in 1968 Pierre Trudeau was elected prime minister, the perception of the federal government about Prince Rupert changed.

Pierre Trudeau, like Prime Minister Wilfrid Laurier who served at the beginning of the 20th century, was interested in developing the trade with Asia to counteract the dependency Canada had developed with the US. His vision was supported by the changing geopolitical and economic situation of the 1970s, with the first OPEC oil crisis that changed the destiny of Prince Rupert. As a result of the surge in oil price, Japanese steel mills cut up to 80 percent of their oil consumption, which they used to import from Middle Eastern countries, and converted to coal, which they started buying from Canada. The prospect that the export of Canadian coal would increase in the years, intertwined with a problem of congestion affecting the port of Vancouver, played in favor of Prince Rupert, which in 1972 was selected as one of the ten ports under the umbrella of the National Harbours Board. Three years later, in 1975, the Fairview Terminal was completed (MacKay 1986: 180–81). Thus, in 1982, CN Rail signed a fifteen-year contract to transport the coal from Prince George to Prince Rupert, which also ensured the overhaul of the tracks to support the increase in traffic to be expected in the years to come. In January 1984, the first cargo of coal left for Japan, and a few months later, the largest dry bulk cargo, transporting 187,000 tons of coal, left from Prince Rupert (MacKay 1986: 192).

In those years, the Macdonald Commission Report was published. In its conclusions, it was stated that "there are a number of reasons to believe that Canada's economic relations with the Asia Pacific rim will continue to strengthen and diversify in the decades ahead [...] The growth in output and trade of these countries is likely to continue to outpace significantly that of Canada's trading partners, the United States and Western Europe" (Macdonald 1985: 254). Prince

Rupert is now at the forefront of the Canadian trade with Asia, with an expanding port that may well be perceived as being a global-scale project in the context of opening new northern areas to development while linking them with key trade regions of the planet, such as Asia (Povoroznyuk et al. 2022: 3).

Building New Infrastructure, Enhancing Connectivity between North America and Asia

New potential shipping routes serve to legitimate development projects, based on potential future trade and on the imagination of global transport and exchange in the Arctic and sub-Arctic regions. In such a context, shortening distances by using the quickest routes, together with modern and efficient means of transport, seems to be a condition of modernity. Profit maximization and efficiency are then invoked and used to shape a narrative and justify the ideology that supports large-scale projects that undeniably favor specific forms of mobilities while rejecting those perceived to hamper development (Povoroznyuk et al. 2022: 5–6). Whereas large-scale infrastructural development generates expectations in terms of future economic opportunities; unexpected economies and alternative development paths may well emerge around possible imagined development and trade, generating hopes for a different and more sustainable development. The anticipatory role of alternative economic paths might then (re)shape and influence the way in which development paths are defined and implemented in a place like Prince Rupert, where the port may well continue to drive the economy, but without phagocytizing other sectors that may be relevant to implement an essential economic diversification that can foster sustainability and sociocultural continuity in the long term.

The port of Prince Rupert hosts five terminals plus a cruise dock, the Northland Cruise Terminal, the only one operated by the Prince Rupert Port Authority. As for the other terminals, they are owned and operated by different companies.[4] Among them, the Fairview terminal has been heavily promoted by the Port Authority as a strength Prince Rupert can offer to Canada (figure 6.2).

Due to the geographic position of the harbor, Prince Rupert offers the shortest way to connect North America to Asia, being between one and two days closer to Pacific Asia than any other port in the Northwestern United States and Canada (figure 6.3). Less time spent on shipping translates into relevant savings for fuel and less risk in shipping operations (Prince Rupert Port Interpretive Centre 2022).

Figure 6.2. *View of the Fairview Terminal from the airport ferry to Prince Rupert, BC, Canada, 2022. © Giuseppe Amatulli.*

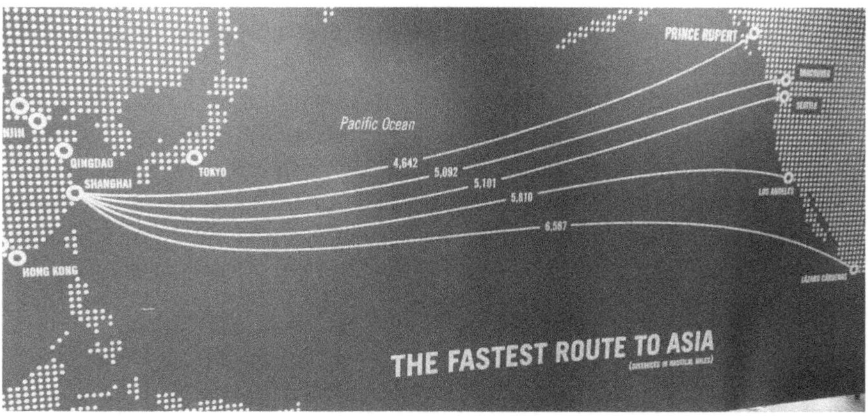

Figure 6.3. *Distances in nautical miles between the Port of Prince Rupert (compared with other North American ports) and Asia. Prince Rupert Port Interpretive Centre, BC, Canada, 2022. © Giuseppe Amatulli.*

Moreover, the port of Prince Rupert, the deepest natural harbor in North America, remains ice-free year-round. This means that it could take on a global leading role in the next few years, when potential Arctic Routes, such as the Northern Sea Route, the Northwest passage and the Transpolar Route may become navigable due to advancements in technology and the receding ice.

Much of the Northern Sea Route is an integral part of the Chinese Polar Silk Road, in which the port of Prince Rupert may serve as a key infrastructure; thus, contributing to re-imagine trade opportunities and possibilities of global connectivity. In the specific case of Prince Rupert, the idea of the expansion of the port shapes future exchange networks while defining new economic and political (im)balances, possible vulnerability, and breakdown points. The port, the biggest infrastructure in the area, is seen as the solution to many economic issues, facilitating "the flow of goods, people, and ideas while allowing for their exchange over space" (Larkin 2013: 328).

Whether jobs and economic growth that the shipping industry generates are often perceived as the main results of the presence of the port, with its expansion that can only bring more opportunities, there are secondary effects that have an impact on other sectors as well. Housing, road expansion, and better social services are expected to be the main issues Prince Rupert needs to face in the years to come. At the same time, local food production (from new forms of agriculture to sustainable local fishing and canning), cultural revitalization of First Nations languages and cultural practices might emerge as a possible alternative to the only economy that has been portrayed as possible so far.

The development of large-scale infrastructures influences the socio-economic organization of a place while shaping the everyday life of people and their needs. The port of Prince Rupert can be seen as an important infrastructure that influences the future paths of the city; however, other alternatives are available and wanted. This was clearly stated by the many inhabitants and First Nation members I met during my stay in Prince Rupert.

My Involvement with the Gitmaxmak'ay Nisga'a Society and the Perspective of Nisga'a Members

A few days after my arrival in Prince Rupert in early July 2022, I attended a canoe meeting organized by the Gitmaxmak'ay Nisga'a Society. I found the advertisement on Facebook, and I was intrigued by it, as the plan was to discuss a five-day canoe trip to take place from July 23 to July 28, 2022. I arrived at the Band Hall at around 7 p.m. and a few members were already there. After introducing myself and asking permission to stay, attendees started to explain the importance of the canoe trip. From the beginning of the conversation, I noticed that a reference to water (and related activities) was made in almost every sentence. Charles, a 65-year-old paddler, gave me the opportunity to make this reflection. When he started talking, he mentioned water in almost every sentence. Fishing and eating, paddling and exploring the outdoors, social and professional life; everything relates to water. For a coastal First Nation, water is a sacred gift, a source of life that connects all living beings. The canoe meeting lasted for around two hours, and it was closed with the decision to host a bingo night the week after to raise money for the upcoming canoe journey. I was invited to join the team and help with the organization of the event.

On Thursday, July 14, I arrived at the Nisga'a Hall in the early afternoon. Charles and Debra had already set up the tables and they were starting to put the auction items on them. Debra quickly introduced me to Sabrina, a lady in her fifties who was double-checking some documents. I asked her how the evening was supposed to unfold and how I could help. She explained that money was supposed to be collected at the canteen (by selling food), through auction and bingo. Then, Debra asked me if I could help set up the bags for the auction items. While doing it, I chatted with Steve and Charles. In his forties, Steve worked for the Nisga'a society. He shared with me some thoughts about Prince Rupert. "You know, there was a lot of fishing here. Almost everyone was involved in fishing. It was good for jobs, opportunities … the community was alive. There were more than twenty bars in town! It was fun … then, everything declined."

According to Steve, the reason why Prince Rupert declined is connected to the fact that everything is outsourced nowadays. In the specific case of fishing, he explained that there are still fishermen around, who get fish in the area. However, due to a lack of processing facilities, once they sell it, fish is sent and processed somewhere else. Steve argued that this is a paradox, as people living in Prince Rupert might end up buying that same fish, but paying much more for it because it was processed somewhere else. At some point during our conversation, Steve's father, Charles, joined us. While drinking coffee, we started talking about the port, its expansion, and the opportunities that it offers to the city. As Charles remarked: "Well, let's see where they are going to expand it. For sure, it is a good thing, in terms of jobs, if they can keep expanding it. However, this island is limited, and there is already much going on. I do not know, I'm not sure this is feasible."

Continuing the conversation, Charles argued that Prince Rupert has the deepest harbor in North America, and this is an advantage that should be used. Then, he explained that he was offered a job at the port, and that he could easily make C$40–45 an hour by working there. Nonetheless, he decided to refuse the offer as working at the port is demanding; long working hours are expected, and it could be stressful for a person's body and mind. Charles decided to keep working as a janitor with the Nisga'a society, making less money, but being in a familiar and friendly environment.

A very interesting perspective on the same issue was provided by Marlena, a young Nisga'a member. In her view, the port brings people away; also because people do not want to work for little money. She moved to Prince Rupert from Terrace, following her husband who is a paramedic. More than jobs and economic opportunities though, she is concerned about cultural continuity in the area. As she shared with me:

> You arrive in Prince Rupert, let's say you get off from the cruise and where are we? Are we represented? Not really! And 50 percent of the population in Prince Rupert is Indigenous. We need to change this. You know, there is a lack of culture here … culture went almost lost. In other areas in BC, in the North, there are so many things happening … cultural camps, gatherings, etc. Here we lack all these things; members do not know … they lost the culture, and they do not feel they belong somewhere.

Marlena strongly believes that facing this issue might contribute to raise awareness about Indigenous cultures and different economic paths that can be followed, not necessarily intertwined with the port. In her argument, she links cultural continuity with economic continuity in a way that might bring relevant changes to how economic activities are conceived in Prince Rupert.

What Marlena meant did not necessarily have (at least for now) an economic goal as its main outcome, but it is mainly directed towards a cultural revitalization goal. As she explained, her dream would be to have a common house in town, with a fireplace where everyone from different communities is welcome and can spend time. People coming from out of town would be welcome there, as a place where they could go and learn about the culture of Indigenous peoples of the area, from Northwestern groups of the coast to the people of the Nass Valley. As remarked on several occasions, she wants to do something to keep the culture alive, and to bring back those features that are on the edge of vanishing. As she told me:

> We still drum, we have traditions, our language ... we need to work to bring everything back. People need to overcome their generational trauma, boarding schools' trauma ... you know, my uncle went to boarding school, the last one was closed in 1994. We need to work to overcome our trauma, and we can do it only by bringing the culture back, without shame!

Salmon Distribution

Cultural practices that have shaped the lifestyle of Coastal First Nations have been defined by water. In particular, salmon has played a key role in shaping the culture while defining communities' livelihood.[5] The importance salmon has until these days clearly emerged during the yearly community salmon distribution event.

I was invited to join the salmon distribution team towards the end of the bingo night that was hosted to raise money for the canoe journey. While speaking with other members, Debra explained that they were expecting to receive a considerable volume of salmon and were in need of volunteers to help with the distribution. I volunteered and was invited to join them at the Nisga'a Hall the following morning.

As I discussed with the fishermen while unloading the fish, salmon distribution is an important practice, a milestone to keep community members together, to keep the sense of identity, the sense of belonging to a community. Salmon distribution is regarded as an important community event, and the Nisga'a government is committed to make sure that members can eat good and nutritious food, locally sourced, as the ancestors used to do. Two trucks with two big boxes of sockeye salmon (500+) arrived at around 1:30 p.m. Unloading it took around one hour, with Steve, Charles, and I working on it. Each household was entitled to get six salmon and people slowly started to line up well before 1:30 p.m. We started the distribution at 2 p.m., and by 4:30 p.m., when the gates closed, most of the salmon was distributed.

While there, I spoke with Charles and Debra about the importance of such an event. As Debra explained, the Nisga'a government pays the fishermen (who are members, too) for their job, and salmon is then distributed to community members for free. What the community receives comes from the Nass Valley, the traditional territory of the Nisga'a. Charles was very excited about the event; he kept saying how good salmon is, the many ways you can cook it, and how delicious it tastes. As he explained: "It's important. People can eat this nutritious and healthy salmon. That's our food! It's important to us that we can eat this kind of food, that we can continue cooking it. You know, we also use to smoke it … but I do not have a rack, otherwise, I would do so."

Salmon is regarded as a powerful fish that can move back and forth between the world of humans and the animal world through the skin, that symbolizes transformation between human and nonhuman forms (Thom 2003: 13). As Charles pointed out, salmon is eaten in many different ways, similarly to the Oolichon,[6] a rich and nutritious fish also known as "The Saviour Fish," because it is the first fish coastal people get after the winter. It is a common resource, and everyone has access to it. That's the Nisga'a philosophy, based on sharing, as everyone who is in need must have access to it.

While distributing salmon, I also had a chat with Jim, a young Nisga'a member, born and raised in Prince Rupert, who wants to move to the south (Vancouver Island or the Okanagan) or to Alberta to work as a mechanic. He clearly said that he does not want to stay in Prince Rupert and work for the port, as he would rather do something that he likes, somewhere else. While sharing his thoughts with me, he commented on the port:

> I have never thought too much about the port, about its expansion. It has always been there, and it is a good thing for Prince Rupert, I guess. But the big expansion … they keep saying that the port is expanding, but I do not see it. I think it would be good, as it will create jobs … but my feeling is that is more like an *expectation*. The expectation has always been about expanding it. But the reality … well, it is more complex.

According to Mason, expectations can be used to build protected spaces that play a key role in the process of supporting technological developments. He defines this specific category of expectations as events collective, which can be seen as a strategic resource for attracting attention from specific sponsors with the aim of stimulating agenda-setting processes and thus contributing to the creation of a specific narrative around the need to have a specific development (Mason 2015: 326–27). Jim pointed out that Prince Rupert is a small town and that it is already remarkable to have such a big port (the third largest in Canada) in such a small town. He said that the expansion might not be as big as the Port Authority wants to depict it. As already pointed out by other residents, Jim remarked that housing is a problem and before expanding the port, the housing situation should be addressed.

On the second day of salmon distribution, I was invited to watch some of the members gutting the salmon. I quickly understood that it was going to be a unique experience, an occasion to see and learn new things, while talking to people and building trust and relationships. As expected, watching was only the beginning, as at some point I was invited to join the members in the activity, under the guidance of an elder, who had been a fisherman for more than fifty years. He showed me how to handle the knife and cut the different parts of the salmon, according to the traditional knowledge members have transmitted from generation to generation. While gutting the salmon, I asked him whether members consume the head of the fish as well. He replied:

> You know, we do not waste any part of the fish. Everything can be used. The head, for example, this is how you should cut it, so you can BBQing it. And you know what? You need to eat the eyes first, as it will make you tune with the fish. So, when you will go fishing the next time, you are tuned with the fish and the fish will come to you, he will give its life to you. And you must be grateful for his sacrifice. This is also valid when you go hunting.

Continuing the conversation, he remarked that in the past people used to fish in an "ocean of fish," as there was abundance. The situation is different nowadays, with the government that keeps setting new rules on how to fish, for instance by defining the net length (Taylor 1999: 4). According to Menzies and Butler, the collapse of salmon stock that took place in the 20th century is due to the different fishing methodologies. Although Indigenous and industrial fishing looked statistically similar, the fishing season was different. Indigenous peoples used to fish various species from March to November while Euro-American settlers preferred to operate between April and June to serve the need of the canning industry (C. Menzies and Butler 2007: 441). According to some Nisga'a elders, the problem is that the government keeps counting how much salmon can be found in a river, without considering that salmon can go back and forth and that other animals (like grizzly bears) can eat it. And bears only eat the skin because it is rich in fat and leave the rest somewhere on the land, which then takes it back, as fish is part of the chain system; its vital components go into the land and the trees absorb it. So, everything is interconnected and interrelated; everything is part of the system and serves to feed nature and the land.

As a First Nation from the coast, it is not a surprise that members have such an attachment to their land and water, and a deep relationship with the fish. Having access to salmon has always had an important meaning for community members; as eating the same food ancestors used to eat, locally sourced, contributes to cultural continuity, in addition to defining people's identity, while making them stay connected to their land and food system, and passing on traditional knowledge and lifestyles (C. Menzies and Butler 2007: 443). Fishing has for a long time been one of the most important ways to make a living in these areas; thus, intertwining cultural continuity with economic self-sufficiency. However, as stated by many Nisga'a members, the lack of fish-processing facilities in the area has seriously affected the ability of people to continue making a living with fishing. Latent fish infrastructure might then be perceived as the reason why economic expectations have not been met. At the same time, it might also provide avenues for individual agency, triggering people's desire to build something different compared to what has been depicted as the only path to follow in terms of socio-economic development.

The Voice of the Northern Native Fishing Corporation

The importance fishing has in a place like Prince Rupert and the side effects the port is having on such an activity was clearly highlighted during a conversation I had with Mabel Mazurek, General Manager of the Northern Native Fishing Corporation. As she told me, fishing is not just a commercial activity, it is much more.

> The port is taking away people from the job and skills from people. Entire families used to fish here! Have you seen that guy who has just entered our office? He is a fisherman, his father was a fisherman, he has three brothers, and they all are fishermen. We used to have people knocking at our doors to ask for licenses, to fish, to keep doing what their father, grandfathers did ... this is slowly vanishing.

Mabel highlighted that it is not possible to keep up with the port, due to the high-paid jobs it offers. She remarked that this is a scary situation because knowledge will be lost. As she pointed out, youngsters are less and less interested in fishing as they want the big money coming from the port, the high paid jobs the port can ensure. That is one of the impacts the port is having on cultural continuity and Indigenous knowledge transmission (Harris 2001: 3) Thus, fishing may well be perceived as much more than just an economic activity, as transmitting the knowledge and fishing skills to new generations is the real meaning of fishing nowadays. In its holistic aspects, fishing may well be seen as an economy of anticipation, that keep asserting Indigenous presence in the area while proposing different economic opportunities that could prevent excessive port development. As Mabel remarked, many Indigenous fishermen do not receive formal education from naval school as they learn everything by doing with their fathers and grandfathers.

The conversation with Mabel was eye-opening. She clearly pointed out how the economic viability and profitability of fishing is intertwined with cultural survival and the transmission of Traditional Knowledge (TK). In fact, when I asked her whether it is still possible to make a living by fishing, she replied:

> It is just impossible. It is not feasible to make a living by fishing. Also because you can only fish salmon for a couple of months per year ... according to the latest regulation, we can put the net in for only 20 minutes and then it is over. Fishing is regulated by the federal government; it is Ottawa that decides what to do and how.

Such a statement highlights a difficult relationship between First Nations and the federal government regarding the management of natural resources. Building Nation to Nation relationships is then fundamental to solve this conflict that is still impacting many bands.[7] Mabel clearly explained how the lack of inclusion in decision-making when it comes to fishing regulation is having a negative impact on Indigenous fishermen of the area. According to Mabel, in 2021, people were not allowed to fish because the government said salmon was scarce; while in 2022, the fishing season ended on August 5, 2022. So, fishing had been allowed for only two months over the summer, and this is a relevant difference compared to the past, where fishing was practiced in different areas and in different periods of the year. This is also the reason why, as Mabel argued, fishermen need a second, sometimes a third job, to get by, as they cannot make a living just by fishing.

Mabel stressed several times that fishing is regulated by the federal government (Meggs 1991: 74-80). It is easy to imagine that fishermen of Prince Rupert are disappointed, to say the least, with the way things are done. Ottawa is perceived as a faraway capital city, and the central government detached from the everyday life of people, unable to understand people's struggles and provide meaningful solutions. Another thing Mabel mentioned was the fact that in the past there were plenty of fishing facilities in Prince Rupert. She said that at least twenty different fishing facilities used to operate in town; nowadays, there is only one facility in Port Edward and one in Prince Rupert, run by CANFISCO—Canadian Fishing Company, owned by Jim Pattison, a multibillionaire who owns different businesses in town.[8] Mabel expressed concern about the fact that CANFISCO has the monopoly of the fishing industry (Gerwing and McDaniels 2006: 260) and facilities could be shut down anytime, thus provoking a devastating impact on local fishermen while posing a further burden for those who are still making a partial living by fishing. As well documented by Menzies, between the 1960s and the 1980s, Prince Rupert was a flourishing fishing town, with several co-op fishers (formed by boat owners and crew members) that owned and controlled the operations of their fishing processing facility and that invested a considerable amount of money throughout those decades (C. R. Menzies 2001: 94–95).

What Mabel shared with me was echoed by Lucy, an elder I met during the July elders' luncheon organized by the Nisga'a society. Lucy pointed out that the port does not create the same number of jobs the town used to have with fishery and the canning industry (Muszynski 1996: 180-181). As she said:

> You know, in the past there were so many jobs, so many people working in the canning industry. Every season was good for something. Crabs, halibut, and salmon in the summer; groundfish in the winter. There was so much to do … fishermen were busy. There were so many hotels, and bars … more than twenty bars in town. Now, there is nothing … it's a ghost town! And you know why? Because everything is sent somewhere else.

Lucy remarked that fishing is still carried out in the area; however, fish is sent to Alaska, where there are processing facilities. She explained she was against the monopoly CANFISCO has created in the last decades, as well as the project of further expanding the port, because those opportunities are not enough to meet the needs of a growing city and its inhabitants while diversifying its economy.

Based on these accounts, it clearly emerges that small and independent workers are impacted by the power dynamics of the market driven economy. If, on the one hand, the port can offer good and high paid jobs, on the other hand it does not provide for everyone while destroying the knowledge people have used for centuries to live and get by. As Mabel remarked towards the end of our conversation: "Fishing has always been here; the skills were important for people to survive. The port might not be here forever, might not provide in the long term what fishing has been able to provide since time immemorial."

The Voice of the Municipality

> It was us or Kitimat; we knew that. And when the possibility of getting LNG [Liquefied Natural Gas] facilities here was concrete, the community got scared. You know, the perspective of receiving 4,000 workers in a few years was not appealing for many of us. The development LNG brings is quick, but the port is different ... slower, I would say.

The discussion about Prince Rupert, the opportunities and potentiality the port can ensure for the city and for the entire community started in this way. Veronika Stewart, Communication Manager of the City of Prince Rupert, shared with me her thoughts on the town, intertwined with what is happening with the port expansion. The port is in constant expansion, with continuous investments that will enhance its capacity, as demonstrated by the investment decision taken in November 2022 by the Minister of Transport to invest C$75 million under the National Trade Corridors Fund to increase the capacity of the Port to export products for green energy and other clean commodities.[9] In Veronika's view, Prince Rupert can grow in a sustainable way; however, the development of the port should be commensurate with the size of the town and the needs of its inhabitants. Otherwise, Prince Rupert will collapse on its own development and dreams to become the biggest port of Canada.

As recent research conducted in North America has showed, resource abundance and/or economic dependence on natural resources has resulted in slower economic growth (Parlee 2015: 425). There is no reason not to believe that this applies to the economic dependence Prince Rupert has been developing with its port. As Veronika pointed out, the possibility that the port of Prince Rupert becomes bigger and bigger, year after year, is real. Nevertheless, the impact such an expansion will have on the city and its inhabitants is yet to be understood.

Whereas the expansion of the port is a huge opportunity for the city, development must be ordered. It is a matter of balance, Veronika highlighted. There is a need to find the right balance to ensure a type of development that the city can support. It is also a matter of providing the right infrastructure the city needs, with roads and housing being the two most significant issues to be addressed in the near future. At the moment, there is not enough accommodation to meet the needs of an expanding port and town, and this is one of the reasons why renting prices have skyrocketed in the past years.[10] On top of that, as Veronika explained, it must be considered that building new houses in Prince Rupert is not easy and very costly due to the high presence of muskeg, without forgetting that the expansion is limited by the presence of a solid block of mountains. Housing is an issue that concerns many residents, conscious that the port and its expansion are having an impact on the city, in terms of sustainability and regarding its socioeconomic development. As one resident told me: "We really have a problem with housing. Rent is so high, as high as in Vancouver! And less than 1 percent of our houses are empty. It really is a problem and if you cannot provide housing solutions, there is no way the city can support any growth."

Talking about these issues related to the expansion of Prince Rupert and its port, as well as the hopes and dreams it generates for the future, Veronika highlighted that the number of jobs the port can generate might be lower than expected. Nevertheless, this is unpredictable for now. What is sure is that opportunities are there, and the city is in a good place to make good use of them while developing in a sustainable way. As she pointed out during our conversation: "In 2001 the pulp mill shut down, in 2007 the port started its expansion thanks to the container terminal. During the six years hiatus, the city got stuck. No revenues, no jobs, people going away. It was quite challenging for the community."

Veronika clearly said that what happened with the mill scared the community, which did not want to have a similar experience with the LNG industry that in early 2000 was thought to become the new source of income for Prince Rupert residents. Many residents were against having huge LNG liquefy facilities in Prince Rupert, mainly because they could not get over what happened with the pulp mill. "What happens when LNG stops? That's what people were demanding and based on the pulp experience, that's why they were opposing LNG. The port is different, firstly because it is diversified.[11]"

Veronika argued that the port can contribute to further diversifying the economy of the town, for two main reasons. First, each terminal trades different resources (propane, wood, coal, and grain, without forgetting the intermodal terminal managed by DP World). It is not like having a huge LNG facility without anything else around. As it has been observed in other parts of Canada, this is an important feature, as a diversified economy can act like a sponge, offering workforce when new opportunities arise while absorbing it when opportunities decline (Usher, Duhaime, and Searles 2003: 177–78). Second, Prince Rupert ensures the quickest way to get to Asia. This is an extremely relevant feature, as it poses Prince Rupert at the forefront of the shipping industry, intertwined with new economic possibilities that may arise in the future. While reflecting on the meaning of these new opportunities, Veronika highlighted that there is awareness about the fact that following this direction will increasingly integrate Prince Rupert in the global market-driven economy. As she explained: "If the market collapses, it's not just Prince Rupert that will go down, everyone will need to deal with that! But the future is unpredictable, nobody knows what is going to happen. Who was able to predict the pandemic? Nobody … so, we cannot really speculate on that!"

Veronika also highlighted how the port is phagocytizing other businesses in town, with huge problems in finding people who want to work in other sectors. It is easy to work for the port and get C$100,000 a year, without having any kind of degree. People do not want to work in other sectors where they will make way less money than that. Thus, cafés, bars, restaurants, and regular shops have trouble finding people. However, working for the port has negative sides as well; a considerable number of workers are on call, meaning that they do not know when they will work and for how many days.

This issue was further explained to me a few weeks later, while volunteering with the Nisga'a society. Dan, a young boy in his late twenties, had been working at the port for four and a half years. He told me the port is expanding, as demonstrated by the recent completion of a new road between Ridley and Prince Rupert, which will cut the truck traffic in town. As he remarked:

> It's going to grow, although I don't know how much, it's difficult to say. They have just had a big hiring round, so I think it is done for the next couple of years now. I am still on call; I do not have shifts. They can call me whenever they need me, and I need to show up. It is difficult to take days off, it's difficult to plan … you must be available to work every time/every day.

Dan explained that there is a list of seniority to get into the union. Once workers are on that list, then they get into the union, and as unionized workers they have more stability and protection. However, as he said, he still needed to wait a bit before being included in that list, as it is extremely unlikely to become unionized before having worked at the port for less than five years.

Obviously, such a situation has detrimental effects on personal life, as well as on community life. Besides not having the possibility to plan how to spend time with family, living with this kind of instability hampers the possibility of people getting involved with the community, doing any kind of volunteering and so finding a way to feel part of the place where they live. Many residents live their life waiting for a phone call from the port, which really compromises the possibility of envisioning a future that is not shaped by the port and its activities.

Residents' Views

"They call it progress … that is what they say it is." (Debra, Nisga'a member)

Promises, intertwined with expectations for a better future, are defining elements of a narrative that promotes economic growth and development (Mason 2015: 339; Flyvbjerg 2016: 11). In a fast-growing city like Prince Rupert, which hosts the third largest port in Canada (after Vancouver and Montreal), this should not be of any surprise.[12] Nevertheless, the lure of jobs and economic opportunities used as leverage by the Port Authority and the city administration to fabricate consent around the expansion of the port does not seem to win the trust of every resident. Such a feeling was highlighted well in a conversation I had with Ken, a university student pursuing a degree in Geography.

I met Ken, a young man in his early twenties, at Cowpuccino's coffee house, on a rainy day. We were queuing for coffee when we started speaking about the town. I explained that I was writing a book chapter on Prince Rupert and its port, and I asked him if there was much going on with the Port Authority. He started explaining that the Port Authority is like an umbrella organization, and that there are different operators, like DP World, Pembina, AltaGas, responsible for running different terminals. The conversation quickly moved to current challenges the growth of the port generates, such as housing and environmental issues. As Ken pointed out:

> There is this plan, Hay 2.0[13] and then the Agenda 2030 for Rupert.[14] You can really see that the plan for the expansion of the city has been done by someone who has just visited Prince Rupert, who is not a local. If you

> are a local, you know certain things, you know what works and what does not for this community.

Ken was skeptical about the fact that growth in Prince Rupert could be sustained at the current level, as reiterated in the latest official community plan (Prince Rupert City Hall 2021: 20).[15] An issue that Ken addressed during our conversation was related to cruise tourism. According to the latest report of the Port Authority, there has been a 230 percent increase in cruise passenger volumes compared to 2019.[16] Such a massive growth in numbers can be explained considering the two-year hiatus due to the COVID-19 pandemic; however, the increase is remarkable and there is no reason to believe that these numbers will drop substantially in the years to come.

Whereas the Port Authority enthusiastically describes the return of cruise tourism, due to the important positive impact on the local economy, with C$ 3.5 million in direct consumer spending in the community, Ken was less enthusiastic about it. He pointed out that if more cruises come to Prince Rupert, the environmental impact will be huge. At the same time, the city center will also suffer as Prince Rupert is not built to support a huge influx of people. Ken argued that planners and politicians might want to copy Victoria when thinking about developing Prince Rupert. However, the two cities are very different and need to address very different issues. As many other people before him, Ken highlighted that one of the most compelling issues in Prince Rupert is housing.

The concern around the housing situation is shared by many I met during my stay in Prince Rupert. No matter the background, occupation, opinions on growth and development paths that Prince Rupert should follow, everybody understands that housing is a big issue that must be addressed. This was well highlighted by Joe, a resident involved with the Urban Farm project, through Ecotrust Canada. As he told me:

> If you think about housing; prices go up because people who work at the port can afford it. So, companies who operate at the port buy old blocks of flats, refurbished them and then rent them out to transient port workers. They can pay C$1,500 per month, even more. This is not a price everyone may be able to afford, especially nonport workers. We have seen these cycles and we are aware of their meaning.

From all these accounts, it emerges that housing clearly is one of the economies surrounding the port and its anticipated (and perpetual) expansion. However, as argued by Joe, sustainability cannot be ensured in this way, especially in a small town like Prince Rupert. As he told me, he had always been interested in food and while the pandemic was still in its first wave, he started volunteering with Ecotrust Canada, which initiated the Urban Farm project in summer 2021.[17]

Ecotrust Canada operates the Urban Farm on a plot of land owned by the Metlakatla Development Corporation, with the aim to raise awareness among residents about the possibilities to produce their own food in their garden.[18] A variety of vegetables are grown during the summer, such as squash, peas, broccoli, tomatoes, fava beans, basil, and kiwi. The farm comprises also a greenhouse, which is, however, not heated, thus not used during the winter. To start the Urban Farm, Ecotrust Canada received C$100,000, which was used to pay for seeds, tools, the rent for the land, and salaries. Although a small project, Joe believes that it can contribute to the debate on food security while revitalizing food knowledge and offering the residents the possibility to get local food and shortening the supply chain.

The Urban Farm project's main goal is to rediscover food traditions and native plants. Raising awareness on traditional food systems is only one of the five priorities (together with climate change, energy, fisheries, indigenous housing) of Ecotrust Canada.[19] To pursue this goal, Ecotrust Prince Rupert is working with schools and First Nations of the area. Workshops with pupils are usually run over the winter season, with the involvement of First Nations, too, which are very much interested in what Ecotrust is doing, as it is perceived as a way to get the traditional food back, together with native plants. This process of getting back and sharing knowledge is also important for future generations, to transmit food knowledge to them.

Towards the end of our conversation, Joe pointed out that eating local food gives the opportunity to reconnect to the land, to traditional knowledge, beliefs, and to a simpler yet fulfilling lifestyle. He also highlighted that there is a need to shorten the food supply chain, as much of the food consumed in Prince Rupert is imported while locally sourced food is exported.[20] For example, fishermen sell fish outside Prince Rupert because they make more money in this way. Fish is profitable and global food companies and restaurant chains can pay much more in comparison to locals. Joe argued that this system is not sustainable and that there is a need to reach a balance, in order to have less import/export of food while providing options to people to meet their food needs with what is available locally.

The work Ecotrust is doing can be then considered as a way to develop an economy of anticipation to respond to the expansion of the port. Producing local food serves to create an alternative economy and a model of sustainability in opposition to the port; potentially drawing tourist revenue, and directly countering the perceived unsustainability of the port, which creates infrastructures that delocalize local resources while favoring the rise of prices in a global market. The Urban Farm is creating new networks to foster food security, while developing infrastructures for the enactment of food sovereignty. New shops selling local food have opened while more and more restaurants are interested in buying locally sourced food to enrich their menu.

This is the case of Fukasaku, a small shop that shares its commercial space with a Japanese restaurant. I read about it on Facebook, and my curiosity was piqued by the advertisement, stating that Fukasaku sells locally produced vegetables. The first time I visited the shop, I left with a bag full of salad, carrots, cucumbers, and zucchini. While filling my bag, I had a conversation with the clerk, who explained to me that most of the vegetables are cultivated in Terrace; however, a small part of them are produced in Prince Rupert too, at the Urban Farm. Fukasaku opened in early 2021, when the pandemic was still on. In spite of it, the shop performed well and there is a sense of pride about the fact that fresh and locally produced food can be offered to Prince Rupert residents. As the clerk highlighted:

> The thing is that in Prince Rupert the narrative is mainly about the port, the big money it brings, the development. I think it is important to have something different as well, and we want to offer some local food to people. Being sustainable means this to us. This could be beneficial for tourism as well, as tourists are curious about our shop; they come and visit us.

Conclusion

"The sun rarely shines in Rupert, but when it does, it makes Rupert the most beautiful place on Earth."

This is a sentence I heard on different occasions during my stay in Prince Rupert. Nisga'a and elderly residents like to describe the city in this way. It is almost as if they want to say that beyond what is visible most days, there is much more, something else that needs to be revealed. Following the same logic, considering the port and its activities, it can be argued that Prince Rupert is not just a port city; it is much more. Although the expansion of the port is advertised as the main development that will bring endless opportunities in the years to come; other activities might ensure a more equitable and sustainable growth. As highlighted in this chapter, different alternative economies are expanding, alternative development paths that are anticipating the port expansion while countering its activity. Local food production, as well as more sustainable fishing, may well be seen as a form of sustainable economic diversification anticipating the pitfalls of port development. Many locals are engaged in promoting a more sustainable approach to development—a balanced development, as many people mentioned. A diversified economy may ensure such a balance, a sustainable development that can bring long-term opportunities. It remains to be seen whether the Port Authority, the Municipality, and the many Multinational Corporations (MNCs) operating the terminals feel the same. For now, it seems that the port tends to phagocytize any different development paths; somehow dictating the destiny of the town and its people.

Already in the 1980s, it became evident that the port provided far less employment than expected. The progressive mechanization, with machines doing most of the work, together with the progressive automation of the terminals, had made it clear that opportunities had to be sought elsewhere (MacKay, 1986: 192–93). Among citizens, there is growing awareness about the need to diversify Prince Rupert and its economy, not to be exclusively reliant on the port and its economic trajectory. During a conversation I had with a councillor of the city of Prince Rupert, a Nisga'a and Metlakatla member, this issue was at the core of our talk. He stated that diversifying economic activities in Prince Rupert is extremely important, as currently the attention is just on the port and many other things are left behind. He would like to see much more happening in terms of agriculture, for example. As he said: "It is not that I would like to shut down the port. It is just that I think it cannot grow forever. There are other things that can be done."

One could argue that Prince Rupert, without an expanding port that can take the lead on the global scene, will not have a prosperous future. Based on what I observed during my fieldwork and the many conversations I had with locals; I argue that it is reasonable to believe that a bright future awaits Prince Rupert, as far as it is not exclusively linked with the continuous expansion of the port and the unsustainable objective of perpetual economic growth. And so, what is the program to make sure that a more balanced and sustainable future can be promoted in Prince Rupert? I would like to answer this question and conclude this chapter by mentioning what was stated during the celebration dinner for the conclusion of the 2022 canoe journey organized by the Nisga'a and other First Nations of the area.

> Every year, before the journey, people ask: What is the program? I always replied: Look at our participants, this is the program! Having this wonderful group of people, ready to go on this healing journey, that is a program in and of itself. When you go on a canoe journey, you leave everything behind, and you gather strength to overcome the difficulties and the challenges you may encounter in your life.

Gathering strength, reconnecting with the ancestors, and getting back to life stronger than ever, ready to face any difficulties. The yearly canoe journey reconnects and heals people, by bringing them back to a time where development was much different in comparison to the current one. As Debra said:

> We do it every year, as a healing journey. We paddle during the day, and we camp at night. That is also the moment when we gather around the fire, and we discuss any issue that might have come out during the day. If anyone has an issue with someone, that is when we discuss it and we handle the issue according to our traditional way, around the fire, as our ancestors used to do.

Perhaps, it is now time to get back together and discuss how to shape the future of Prince Rupert. Stakeholders and rightsholders, public officers and citizens, First Nation members and settlers living in the area, should come together to define how they envision the future, what they would like to see, and what they do not want, while delineating alternative development paths that can ensure the city a brighter and more sustainable future.

Giuseppe Amatulli is a postdoctoral fellow in the Rebuilding First Nations Governance (RFNG) Research Project, at Carleton University. His research methodology, anchored in a strong community-based approach, allows him to perform cutting-edge qualitative research at the intersection of socio-legal and environmental anthropology, intertwined with Indigenous Rights and Indigenous Governance. Giuseppe obtained his PhD in Anthropology from Durham University (2023), and before joining Carleton he was a visiting postdoctoral researcher at Tromsø University in the Arctic Silk Road Research

Group, in the context of which he conducted fieldwork in Prince Rupert, BC, in the summer of 2022.

Notes

1. Excerpt of the conversation I had with a taxi driver once I arrived in Prince Rupert on Monday, July 4, 2022.
2. Chiang Chuck, February 5, 2021, 'Prince Rupert aiming for top port status.' https://www.biv.com/news/transportation/prince-rupert-aiming-top-port-status-8263196 (last accessed on February 25, 2025).
3. Prince Rupert Port Authority, February 24, 2022, 'Proposed Second Terminal in Prince Rupert.' https://www.rupertport.com/active_project/secondterminal/ (last accessed on February 6, 2023).
4. The Ridley Island Propane Export Terminal is a Bulk Liquid Propane Terminal operated by AltaGas Ltd. The Trigon Pacific Terminal, a coal terminal with a capacity to load up to 9000 tons per hour, is operated by Trigon Pacific Terminals Ltd. The Westview Wood Pellet Terminal, a wood pellet export facility with a capacity to ship 1.25 million tons of wood pellets per year, is operated by Drax. The Prince Rupert Grain Terminal, one of the most modern grain facilities in Canada, ensures a capacity of shipping up to 7 million tons of grain annually, operated by Prince Rupert Grain. The Fairview Terminal, the first dedicated intermodal container terminal in North America, is operated by DP World. Source https://www.rupertport.com/terminals/ (last accessed on September 7, 2022).
5. Laiwan, April 20, 2015, 'Return to the water: First Nations relations with salmon.' http://www2.laiwanette.net/fountain/return-to-the-water-first-nations-relations-with-salmon/ (last accessed on December 15, 2022).
6. Nyce Sr., Harry & Desson, Edward, July 10, 2018, 'The Saviour Fish: Protecting Nisga'a Connection to Oolichan.' https://coastfunds.ca/stories/the-saviour-fish-protecting-nisgaa-connection-to-oolichan/ (last accessed on February 25, 2025).
7. Clark, Ian, February 11, 2017, 'Nation-to-Nation Relationship.' http://www.atlas101.ca/pm/concepts/nation-to-nation-relationship/ (last accessed on December 16, 2022).
8. Canfisco Group, https://canfisco.com/about-canfisco (last accessed on December 19, 2022).
9. Transport Canada, November 16, 2022, News release, 'Government of Canada invests in the Port of Prince Rupert to increase supply chain capacity.' https://www.canada.ca/en/transport-canada/news/2022/11/government-of-canada-invests-in-the-port-of-prince-rupert-to-increase-supply-chain-capacity.html (last accessed on February 25, 2025).
10. Bartlett, Keili, December 15, 2017, 'Rent continues to rise in Prince Rupert, drops in Terrace.' https://www.haidagwaiiobserver.com/news/rent-continues-to-rise-in-prince-rupert-drops-in-terrace-6335656 (last accessed on December 13, 2022).
11. Orton, Tyler, June 1, 2014, 'Port and energy boom reignites Prince Rupert's economy.' https://biv.com/article/2014/06/port-and-energy-boom-reignites-prince-ruperts-econ (last accessed on December 13, 2022).

12. Chiang, Chuck, February 8, 2021, 'Port of Prince Rupert could soon trump Montreal.' https://www.westerninvestor.com/british-columbia/port-of-prince-rupert-could-soon-trump-montreal-3832751 (last accessed on December 19, 2022).
13. The City of Prince Rupert, November 2015, 'Hays 2.0 – A Modern Vision for Prince Rupert.' https://discoursemedia.org/uploads/Hays-2-Vision.pdf (last accessed on February 25, 2022).
14. Beasley & Associates, Planning Inc., CIVITAS Studio – Urban Design and Planning/CIVITAS Architecture, December 2019, 'Prince Rupert 2030, The Vision. A Vibrant City Hosting a Vibrant Port.' https://static1.squarespace.com/static/56953d59d82d5e4a6a3ee499/t/5df80aba6eabde6fa227a53d/1576536893173/191206_PRU_VisionReport_web.pdf (last accessed on December 19, 2022).
15. The City of Prince Rupert, 2021, 'Official Community Plan.' https://www.princerupert.ca/building-development/official-community-plan (last accessed on February 18, 2025).
16. Prince Rupert Port Authority, October 26, 2022, 'Strong return for cruise tourism in Prince Rupert.' https://www.rupertport.com/strong-return-for-cruise-tourism-in-prince-rupert/ (last accessed on December 19, 2022).
17. The Skeena, August 17, 2022, 'Check out the fresh and delicious food grown at this urban farm in Prince Rupert.' https://www.theskeena.com/2022/08/17/check-out-the-fresh-and-delicious-food-grown-at-this-urban-farm-in-prince-rupert/ (last accessed on December 20, 2022).
18. Millar, Kimberley, March 24, 2021, 'Urban agriculture area planned for Prince Rupert downtown core.' https://www.thenorthernview.com/news/urban-agriculture-area-planned-for-prince-rupert-downtown-core/ (last accessed on December 20, 2022).
19. Ecotrust Canada, 'Our Approach.' https://ecotrust.ca/approach/ (last accessed on December 20, 2022).
20. Fawcett-Atkinson, Marc, November 19, 2020, 'Building a farm – and food security – where a northern highway ends. https://www.nationalobserver.com/2020/11/19/news/prince-rupert-farm-food-security-COVID-19 (last accessed on February 6, 2023).

References

Flyvbjerg, Bent. 2016. "Introduction: The Iron Law of Megaproject Management." In *The Oxford Handbook of Megaproject Management*, ed. Bent Flyvbjerg, 1–18. Oxford: Oxford University Press.

Gerwing, Kira, and Timothy McDaniels. 2006. "Listening to the Salmon People: Coastal First Nations' Objectives Regarding Salmon Aquaculture in British Columbia." *Society and Natural Resources* 19(3): 259–73.

Harris, Douglas C. 2001. *Fish, Law, and Colonialism. The Legal Capture of Salmon in British Columbia*. Toronto: University of Toronto Press.

Larkin, Brian. 2013. "The Politics and Poetics of Infrastructure." *Annual Review of Anthropology* 42: 327–43.

Macdonald, Donald. 1985. "*Macdonald Report—Canadian Opportunities in an Open World, Part II.*" *Report of the Royal Commission on the Economic Union and Development Prospects for Canada*. Ottawa: Minister of Supply and Services Canada.

MacKay, Donald. 1986. *The Asian Dream: The Pacific Rim and Canada's National Railway*. Vancouver: Douglas & McIntyre.

Mason, Arthur. 2015. "The Social Life of a Promise-Disappointment Cycle." In *Subterranean Estates: Life Worlds of Oil and Gas*, ed. Hannah Appel, Arthur Mason, and Michael Watts, 325–39. Ithaca: Cornell University Press.

Meggs, Geoff. 1991. *Salmon: The Decline of the British Columbia Fishery*. Vancouver: Douglas & McIntyre.

Menzies, Charles R. 2001. "Us and Them: The Prince Rupert Fishermen's Co-Op and Organized Labour, 1931–1989." *Labour/Le Travail* 48(48): 89–108.

Menzies, Charles, and Caroline Butler. 2007. "Returning to Selective Fishing through Indigenous Fisheries Knowledge: The Example of K'moda, Gitxaała Territory." *American Indian Quarterly* 31(3): 441–64.

Muszynski, Alicja. 1996. *Cheap Wage Labour: Race and Gender in the Fisheries of British Columbia*. McGill-Queen's University Press.

Parlee, Brenda L. 2015. "Avoiding the Resource Curse: Indigenous Communities and Canada's Oil Sands." *World Development* 74: 425–36.

Povoroznyuk, Olga, et al. 2022. "Arctic Roads and Railways: Social and Environmental Consequences of Transport Infrastructure in the Circumpolar North." *Arctic Science*: 1–94.

Prince Rupert City Hall. 2021. "Official Community Plan." Retrieved April 16, 2025 from https://www.princerupert.ca/building-development/official-community-plan.

Prince Rupert Port Interpretative Centre. 2022. "Canada's Leading Edge", Visitor Information Centre, Prince Rupert, July 2022.

Taylor, Joseph E. 1999. *Making Salmon: An Environmental History of the Northwest Fisheries Crisis*. Seattle: University of Washington Press.

Thom, Brian. 2003. "The Anthropology of Northwest Coast Oral Traditions." *Arctic Anthropology* 40(1): 1–28.

Usher, Peter, Gérard Duhaime, and Edmund Searles. 2003. "The Household as an Economic Unit in Arctic Aboriginal Communities, and Its Measurement by Means of a Comprehensive Survey." *Social Indicators Research* 61(February): 175–202.

Wilson, Gary N., and Paul Bowles. 2016. *Resource Communities in a Globalizing Region: Development, Agency, and Contestation in Northern British Columbia*. Vancouver: UBC Press.

CHAPTER 7

The Promises and Fears of Infrastructure

An Alaskan Port Expansion, Geopolitics, and Local (Dis-)Engagement

Olga Povoroznyuk and Peter Schweitzer

> By most of the laws of nature, Nome should never have been the site for a port city. There was no safe harbor for ships at the mouth of the Snake River, and it was dangerous to land or to take a small boat inside the mouth of the river. Large ships would be forced to anchor several miles off-shore and unload their passengers and freight to lighters and shallow-draft barges that could be run up on the beach. (Cole 1984: 29)

Notwithstanding Terrence Cole's observation about the unsuitability of Nome as a port city, on December 21, 2020, the US Congress authorized a deep-water port in Nome, Alaska (Rubano 2021). The project was included in the 2020 Water Resources Development Act (part of the Consolidated Appropriations Act of 2021) and consisted of extending and existing causeway and of dredging a new deep-water basin at the port. This decision came after a lengthy process of determining the "optimal Arctic port" for the US (VanderBerg 2018) and years of waiting. While some community leaders, including the mayor of Nome, the port director, and the CEO of the Bering Straits Native Corporation, lauded the decision and pointed to the expected economic benefits, others—such as the nonprofit regional tribal consortium Kawerak—raised concerns regarding subsistence and environmental impacts (Rubano 2021).

The enormity of the project, including a price tag of US$500+ million, means that—despite congressional authorization—funding is only secured for part of the project (Milligan 2022). The extent of the endeavor might also lead to the assumption that the local population of this relatively small settlement (approximately 3,500 residents) is either excited by the "promise of infrastructure" (Anand et al. 2018) or up in arms about the potential dangers to the community and its environments, that is about the "infrastructural violence" (Rodgers and O'Neill 2012) involved. Instead, the predominant attitude that we encountered during preliminary fieldwork in Nome in July 2022 was one of disinterest and disbelief that the project would make a difference, neither for better nor for worse. Having worked ourselves in the past with large projects that bore the characteristics of the "enchantments of infrastructure" (Harvey and Knox 2012) or have been perceived as "transformative infrastructure" (Povoroznyuk 2022) by local residents and external experts alike, the situation we encountered in Nome was somewhat unexpected.

In the remainder of this article, we will try to understand this specific constellation. We will do so by exploring a variety of past and present processes within the community and beyond, and by putting them into conversation with what our interlocutors told us in the summer of 2022. The starting point will be a brief overview of the history of Nome, as well as of its port, in order to understand the social, political, spatial, and demographic dimensions of the community's past and present. This will be accompanied by analyzing relations across Bering Strait over time, given that this maritime passage is a chokepoint (Carse et al. 2018) along the Polar Silk Road (see Introduction), and the often-volatile relations between Russia or the Soviet Union and the United States of America are an important ingredient for local dreams, fears, and expectations. These dynamics will be portrayed in order to contextualize local perceptions of the Nome port expansion plans. The final section will attempt to make sense of these intersecting and, at times, contradictory vectors of development, stagnation, and anticipation.

History of Nome and Its Transport Infrastructure

Returning to Terrence Cole's quote above, one is tempted to ask why a site seemingly ill-suited for safely landing small boats and for large ships to anchor nearby was selected for a place that has become the regional hub of the Seward Peninsula. The answer is simple: the beaches of Nome and the creeks nearby were full of gold and thus attracted a large number of fortune seekers at the end of the 19th century. Thus, the official (non-Indigenous) narrative of the history of Nome sets the year 1898 as the temporal starting point. According to this story line, the "three lucky Swedes"—John Brynteson, Eric Lindblom, and Jafet Lindeberg (the latter was actually Norwegian—Haycox 2002: 206)—struck gold on the Snake River and its tributaries in September 1898 and staked their claims in October of the same year (Cole 1984: 11–27). It took until 2005, when the Snake River Sandspit Late Western Thule archaeological site was discovered during port improvement works (Eldridge 2014), that Indigenous presence before the gold rush was being acknowledged. This dichotomy between the public image of Nome as a gold rush town and the less well-documented Indigenous presence in the region continues to a certain degree up to the present day. Among other things, this is illustrated on Anvil City Square, at the very center of Nome, where statues of the "three lucky Swedes", as well as a giant gold pan, had been providing a rather one-sided take on history. Only in 2010, an additional statue representing two Inupiaq boys, who according to oral tradition helped the Swedes to find the gold, was added to Anvil City Square.

The 1898 discovery set off a proverbial gold rush, bringing the population of Nome, which was called Anvil City back then, to an all-time high of 20,000 in 1900, making it the largest town in Alaska at that point in time. The population started to dwindle after that, due to the depletion of the non-renewable resource, and fell to 852 in 1920 (that is, during the Spanish influenza pandemic). The number of residents has been increasing again ever since then, with the largest gains occurring during the 1920s and 1980s. Since 1990, the population development has stagnated at a level of about 3,500+. The percentage of Alaska Natives has increased over time, however, currently constituting 58 percent of Nome's residents (Crouch and Lewis 2023: 94). Thus, one could say that the settler community of Nome has never really recovered from the post-gold rush decline; the Indigenous community has been thriving demographically. One must not forget, however, that severe discrimination and segregation characterized living conditions for Alaska Natives in Nome for the better part of the 20th century (Cole 2001).

Transport and communication and transportation infrastructures—even across national boundaries—were quickly created after the discovery of gold in Nome in 1898. In 1899, a mail route from Dawson City (Canada), the hub of the Klondike gold rush area, to Nome was established (Schneider 2012: 113). That is, a man named Ben Downing was issued a US Postal Service contract to deliver the mail from Dawson to Nome by dog team (Schneider 2012: 5, 27). Starting in 1900, the Washington and Alaska Military Cable and Telegraph System (WAMCATS) was constructed by the US Signal Service and connected Nome to other parts of Alaska and beyond. While initially built as a military communication system, it provided the trail infrastructure for dog team mail carriers and other travelers (Schneider 2012: 10–11).

Interestingly, debates over the port of Nome are basically as old as the gold rush settlement itself. A request to establish Nome as a port of entry, which would have promised economic prosperity, was denied by the US Treasury Department in 1900/1901 (Jones 2007: 40–42). Even the construction of a one-mile-long deep-water pier extending into Bering Sea was contemplated as early as 1900, and the *Nome News* determined the development of harbor facilities as critical for the future of the town (Jones 2007: 43–44).

An expanded port, possibly with the designation "port of entry", was not the only transport infrastructure tied to expectations of a bright future for Nome. Plans for railroads, either with regional aspirations or with the promise of connecting to other railroads, were abundant during the early days of Nome. One of the most adventurous plans in that respect has been to build a railroad bridge or tunnel across Bering Strait to connect North American and Eurasian rail systems. The first such plans appeared in the 1880s (Cole 1990: 3–4), thus preceding the establishment of Nome. What first was a quite unrealistic proposal for a bridge crossing Bering Strait became a tunnel project that has generated hope and hype in Nome and beyond periodically. The latest testimony to the power of that dream is a documentary film entitled *The Strait Guys* that was released in 2022 (Minnich 2022).

Racial discrimination has a long history in Nome, as it has elsewhere in the United States. Bathsheba Demuth (2019: 212) states, referring to the gold rush days, "Nome was essentially a segregated town." For a later period, Terrence Cole 2001 [1992]) has pointed to the existence of "Jim Crow policies" in Alaska and has used Nome as one of its examples. One infamous incident occurred in Nome in 1944 and involved Alberta Schenck, the daughter of an Inupiaq mother and a white father (see also Rogers 1998 for a short Indigenous commentary on the incident). She worked as an usher at Nome's only movie theater, where she saw racial segregation in practice: "Natives" and "half-breeds" had to sit on the right side of the theater, while only

"Whites" were allowed on the left, no matter how crowded the right side became. Alberta complained about the situation, was fired, and wrote a letter to the local newspaper, *The Nome Nuggett*, criticizing these policies as not being in the spirit of the US Constitution. Soon after the letter had been published, Alberta went on a date with a white army sergeant to that very Nome movie theater. After she and her date refused to sit on the right side, she was arrested and had to spend a night in jail. The uproar and protests following the letter and the arrest led eventually to the passage of the Alaska Equal Rights Act of 1945 (Cole 2001 [1992]). Still, as many residents and observers would confirm, discrimination did not end with 1945.

Summarizing the historical trajectory of Nome over the last 125 years, one is tempted to conclude that the initial boom that brought Nome as we know today into being was never followed by a comparably large economic boost. At the same time, promises of a grand future—based on resource extraction and/or infrastructural development—have been reoccurring periodically. It seems that these promises were primarily geared toward the settler population and paid little attention to Indigenous imaginations.

Figure 7.1. *Wooden installation "Welcome to Nome." © Olga Povoroznyuk.*

Nome as Part of the Bering Strait Region and the Wider Geopolitical Context

Nome and the Seward Peninsula, that is, the US American areas closest to the Russian Federation in terms of physical space, are part of Beringia, which is defined as the land and sea between the Lena River in the west and the Mackenzie River in the east. The historical significance of Beringia lies in the fact that it formed the "Bering Land Bridge" (Hopkins 1967) between Eurasia and the Americas during periods of maximal glaciation, including during the last ice age, when the land bridge provided a corridor for the peopling of the Americas. Consequently, the different parts of Beringia have been affected by related prehistoric human migrations and cultural traditions (see, e.g., West 1996; Hoffecker and Elias 2007), and characterized by similarities in flora and fauna in ecosystems that have changed radically over the millennia (see, e.g., Hulten 1937; Romanenko 1994; Abbott and Brochmann 2003). The shared but divided fates of the people living on both sides of the Bering Strait have recently been aptly joined in an "environmental history" focusing on the 19th and 20th centuries (Demuth 2019).

Human contact and exchange did not stop with the flooding of the land bridge and the emergence of Bering Strait as we know it today. While these connections had been a major focus of the Jesup North Pacific Expedition at the turn of the 19th to the 20th century (Boas 1905; Kendall and Krupnik 2003), they did not receive much attention during the Cold War. Not surprisingly, these connections were "re-discovered"—often under the label of "crossroads of continents" (Chaussonnet 1995; Johnson, Hickey and Hoover 1991; Fitzhugh and Crowell 1988)—during the late 1980s and early 1990s, when the Cold War seemed to have ended. Indigenous contacts across Bering Strait lasted all the way to 1948 (Schweitzer and Golovko 1995; Schweitzer 1997), when the iron curtain (or ice curtain) came down in that part of the world as well.

Diplomatic relations between the United States and the young Soviet Union did not start until 1933. During World War II, however, Nome began to play an important role in contacts across Bering Strait again, as part of the Lend-Lease program (Hays 1996). Between 1941 and 1945, approximately 8,000 aircraft and supplies were brought to Russia via Nome. The above-mentioned Cold War, however, ended friendly and regular contacts for more than four decades, except for the odd clandestine contact among Indigenous hunters (Bruemmer 1977). The silence lasted until 1988, when a so-called friendship flight between Nome and Provideniya re-established contacts (Ramseur 2017). An outcome of these renewed bilateral contacts was the signing of an agreement between the United States and the USSR in 1989—which entered into force in 1991—to allow visa-free travel for Indigenous residents to visit relatives on both sides of Bering Strait. This agreement has remained in effect, and was reinstated in 2015, after it had been curtailed between 2012 and 2015 (Russell 2015). Still, the number of people making use of the agreement decreased over the years compared to the heydays of the early and mid-1990s.

While the Northern Sea Route (NSR), the passage along the Arctic coast of the Soviet Union, became a logistical reality in the 1930s, regular usage had to wait until the 1960s. As the NSR largely remained an internal shipping route during Soviet times, mostly used to supply Arctic communities and to ship out resources from there (see Chapter 2 for a detailed discussion of this configuration), the impact on Bering Strait and the Seward Peninsula was minimal. Although the NSR opened up to foreign ships in 1991, the 1990s did not see any increase in shipping activities due to socioeconomic turmoil resulting from the dissolution of the Soviet Union. During the 21st century, Russian and foreign transits of the NSR started to increase, including LNG shipments from Sabetta, which increasingly worried the residents of the Seward Peninsula. In the years before the Russian attack on Ukraine, various scientifically based suggestions for safe and culturally and ecologically sound shipping in the Bering Strait were made, which addressed Russian traffic as well (Huntington et al. 2015, 2019). While the Corona pandemic resulted in a slowdown in 2020 and 2021, the Ukraine war, and the sanctions it triggered, led to a collapse of foreign shipping along the NSR (Humpert 2022). Nevertheless, the Russian Federation remains upbeat about the prospects of the NSR: in early 2023, Rosatom, the agency responsible for the passage, forecasted to ship 270 million tons through the NSR, which would be almost ten times the amount transported in 2022 (Humpert 2023).

The port of Nome has never been used by ships transiting the NSR and, given the current geopolitical tensions, this will not be the case in the foreseeable future. Thus, for Nome, the only option of servicing ships traveling from the Atlantic to the Pacific Ocean, or vice versa, would be ones using the Northwest Passage (NWP). Given the current tense relations between the Russian Federation and other Arctic states, as well as the notoriously difficult passage of the NWP, it is unlikely that the NWP will contribute a lot of business to the expanded port of Nome. The Transpolar Sea Route, however, might become a reality by the middle of the 21st century if sea ice decline in the Arctic Ocean continues (Bennett et al. 2020; Huntington et al. 2023). This would open new possibilities for a deepwater port near Bering Strait. Interestingly, despite many publications devoted to Arctic shipping in recent years, Arctic ports—the land-based infrastructure it requires—have not seen the same amount of scholarly attention as shipping itself (Panahi et al. 2021).

Local Perceptions of Port Expansion Plans

Our conversations in Nome and surroundings involved local "experts," primarily decision-makers, and researchers, as well as "regular residents" (Indigenous people and organizations and non-Indigenous settlers). The main selling points of the port project had been articulated primarily by experts and focused on aspects of economic growth, such as increased business opportunities and more jobs, reduced consumer prices, growth in the tourism sector, and providing services to cruise ships, oil tankers, fuel barges, research ships and icebreakers alike. In addition, geopolitical security concerns, spanning local, and especially national and international levels, were occasionally mentioned as justification for the project. For example, one official person responsible for the port extension didn't see any negative impacts relating to the project as a "child" that had been nurtured for the last several years. The citation from the interview below is filled with expectations of development and opportunities and asserts the importance and timeliness of the new port:

> You know we are just barely coming out of a recession here. Everything that happens here is a delay of what happens outside, so we are just coming up and out of it ... We didn't see tour ships, we didn't see as many research vessels, but the cargo and the fuel, and the gravel have to move! So, we moved ... But we are nowhere near where we need to be, where we used to be without economic boom, and we need one! The region needs one, it needs jobs![1]

Likewise, the current mayor of Nome, John Handeland, is also in support of the project, as was to be expected. In our conversations with him, he pointed out that Nome had been selected as the first (and, so far, only) Arctic port of the US, which understandably made him proud as a city leader. At the Arctic Encounter 2023 meeting, he presented himself as the mayor who managed to secure vital funding for the port through a once-in-a-generation federal spending package (Monet 2023). Given that similar plans had been discussed for close to half a century, his remark that "... most people thought it was a pipe dream before we were fortunate to get funding" (Monet 2023) needs to be understood in this historical context. At the same time, he acknowledged potential environmental concerns affecting sea mammals and other resources important for subsistence activities. At the Arctic Encounter 2023 symposium, he said in reference to critical voices, "I'm not belittling people for having those feelings," and "...I'm not white-washing the past ..." (Monet 2023). Still, the question remains whether this kind of understanding is sufficient for addressing the concerns of critics.

The Port Commission was established as an advisory body to the City Council of Nome. It consists of seven experts—marine biologists, historians, politicians, and local stakeholders, having a background in maritime industries and transport infrastructure. Since its foundation, it has served as a mediator between local decision makers, the port administration, and the community. Such a role obliges the commission members not only to be knowledgeable about the local issues and to listen to the people's concerns but also to look for possible solutions. While two of the commission members with whom we spoke during our fieldwork in Nome had a comprehensive understanding of the potential consequences the port expansion might trigger, their individual evaluations of the project differed significantly. One of the experts had a rather conservationist point of view and saw a lot of potential negative impacts on the habitat of marine mammals and subsistence activities due to increased shipping. However, the other one seemed to have a pro-development stance and a more optimistic outlook based on the expectation that an influx of construction workers and military personnel will stimulate consumption and economic growth.

A few other local officials and public figures highlighted similar issues and argued that the project will strengthen national security and stimulate economic development. Scientists, on the other hand, pointed to the cumulative effects of unregulated marine traffic coupled with climate change impacts (Kettle et al. 2020). They argue that the new sea port and increased shipping traffic coming with it, will impact habitats and the behavior of marine mammals and fish—important subsistence resources for Indigenous and other local residents of Nome. On a larger scale, competition for commercial fish resources (such as pollock) between the Alaskan and Russian sides of Bering Strait is likely to increase in this new situation.

A much longer list of environmental and other concerns connected with the port extension project was voiced by Indigenous residents of Nome. A community meeting with elders, advocates, and leaders of major native corporations represented in town, revealed specific environmental, subsistence, health and economic concerns associated with the new infrastructure project. Remarkably, most of them were formulated in the (post-)colonial context, and were linked to sentiments about having too little voice in such important community affairs as the port planning. It was pointed out that Indigenous peoples are still underrepresented in town, evidenced by the fact that the archaeological artifacts proving Indigenous habitation of the area were only recently recognized and exhibited in the local museum. With much delay, statues of Indigenous hunters were finally erected on the main city square (Anvil Square) in addition to the "three lucky Swedes," whose monument had been standing there for some time. Some felt that the City of Nome and the State of Alaska focused too much on issues of national security, thereby pushing aside concerns over environmental and food security, climate change impacts, or suicide rates. There was a consensus that Arctic shipping will have adverse impacts on marine habitats. As commercial fishing in the Chukchi Sea is still unregulated, negative impacts are to be expected. A specific concern was raised about possible environmental degradation and pollution coming from Russian LNG tankers. There was also fundamental opposition to such port expansion plans in general, which could be summarized in the sentence "we do not want infrastructural development." Finally, we were also told that new transport infrastructures can cause sickness and starvation.

A conversation with a representative of the Indigenous corporation "Kawerak" who is very familiar with the sentiments of local and Native residents regarding the port expansion project and concerns about the future of subsistence activities in the Nome area made some of these connections quite clear:

> I think people have a lot of questions about the port expansion, what it means for Nome ... What does it mean in terms of jobs? Is it actually going to create long-term jobs for people? If so, where are those people going to live because Nome's housing situation is critical? ... The environmental concerns have to do with habitat concerns. You know the Nome region has had the only Tier II salmon fishery in the whole state of Alaska in its history. And a Tier II fishery is like ... extremely restricted fishery. You have to prove you have historically used and you have to meet certain criteria. That was in late 1990s and early 2000s. No one really knows what caused it. There are lots of theories and there are probably

> lots of factors but, you know, environmental degradation certainly is one potential cause. There's concern about our salmon populations, people hunt walrus at the port of Nome—it's a hunting location—and there's been a lot of controversy about that between hunters and the city, and between tribes and the city.[2]

Indigenous leader and activist A. B., who was also part of our community meeting with Indigenous residents of Nome, is known as one of the most outspoken critics of the project. While we did not have a chance to talk with him in a detail after our introductory meeting, a recent interview he gave a journalist provides some background to his opposition to port development: "B. said that his memories and the legacy of the Iñupiat who have lived here for thousands of years shouldn't have to compete with the modern-day monetary gain some people hope to capitalize on as the Arctic becomes increasingly ice-free" (Schwing 2023).

The list of concerns among non-Indigenous residents was slightly shorter than for Indigenous inhabitants. A dominant concern was the lack of housing and other community infrastructures that would be further exacerbated by the port expansion. Rising housing prices already had negatively impacted many local people, and an anticipated major population influx would make things worse. Nontransparent investment practices and the potential for corruption surrounding port construction were mentioned as well, as was an increase in crime more generally. Environmental concerns such as pollution and disruption of marine habitat by vessels were also voiced. The expanded port will limit access to subsistence resources (such as salmon, walrus, or whales), and expose the community and region to oil spill risks. Some local interlocutors also thought that national security concerns, which might lead to the remilitarization of the Seward Peninsula, predominate over the economic rationale of the port expansion. The new port will not have a positive impact on community supplies, as groceries and goods are increasingly being delivered by air. Proposed benefits, such as part-time or seasonal jobs or the construction of camps to house workers will be temporary, while increased opportunities for selling souvenirs to tourists are being overshadowed by concerns about pollution from these very cruise ships and the lack of infrastructure and services that are necessary to accommodate mass tourism.

Another question is how realistic it is to assume that the full funding that is necessary for the completion of the project will be forthcoming. At a recent Nome City Council Common meeting (April

2023), the port director provided an update about the permitting process that was overall optimistic and upbeat (Knight 2023). Still, in the process it became clear that there are several open questions, for example the cost-sharing agreement between the federal and the local government, which is supposed to be 90/10 in favor of the local purse. Harbormaster Baker, however, pointed out that "… there is no concern about getting there because the 90/10 cost share, under the language that was passed in the National Defense Authorization Act, has become law" (Knight 2023). According to the recent talk by Mayor Handeland at the Arctic Encounter symposium in March 2023, no funding agreement was signed by the City of Nome at that point; however, the port expansion plans remained in force (Monet 2023).

One of the reasons for that is the new geopolitical situation Nome finds itself in since the beginning of Russia's invasion of Ukraine. While the Nome port expansion was decided before the war, and also before President Biden's infrastructure investment package, the "new Cold War" makes the Bering Strait a potential conflict zone between superpowers on a confrontation course. After our stay in Nome, Senators Murkowski and King introduced the Arctic Commitment Act to Congress in August 2022, which would establish a permanent maritime presence in the US Arctic and implement Arctic Strategies at the Port of Nome (Arctic Commitment Act 2022). While the Act has not yet passed Congress, the introduction of the bill seemingly surprised Nome and the Bering Strait region (Loewi 2022). According to Handeland, a deep draft port would allow US Coast Guard icebreakers and Navy warships to be docked there. He also mentioned Nome's historical connection with the military, ranging from the Alaska Territorial Guard to World War II, the Cold War, and both Desert Storm operations (Monet 2023). Some experts and local residents we talked to also thought that the port expansion will be part of a larger process of remilitarizing the region and the country vis-à-vis its neighbor across the Bering Strait.

If we try to summarize the prevailing types of local attitudes, we can see a variety of positions. While some of our interlocutors voiced a number of concerns as presented above, several others expressed moderate expectations that nevertheless did not sound very optimistic. However, in most cases, local residents seemed to have split opinions: while seeing potential positive impacts of the expansion project, they were concerned with its multiple negative consequences. Most surprising to us was the large number of people who showed no real interest in this mega-project. While some argued that this was due to a lack of information and involvement in decision making about the project, others claimed preoccupation with other, more pressing issues. Interestingly, another opinion voiced was that the port expansion alone will not be sufficient to trigger positive change for Nome without a clear economic rationale and investments into social projects (such as investments into permanent housing).[3] In short, a town built on the more or less constant anticipation of a boom, seemed not to be "enchanted" by the prospect of an expanded port, which would mean extensive construction activities, as well as bring money to other businesses in town.

Generally, it seems that most of our interlocutors were preoccupied with the main unanswered question: how will the port expansion address the most critical social issues of Nome, such as the lack of affordable and quality housing, high consumer prices, homelessness, substance abuse, as well as high rates of suicide and crime? Or, in short, why do local residents in Nome need a bigger port? In the wake of large-scale projects, such legitimate questions and concerns of residents are often being drowned out by the loud voices of development boosters and advocates of progress. Therefore, the question that we asked while analyzing our fieldwork data on local perceptions of the port expansion project was: were local residents vocal enough

Figure 7.2. *The Port of Nome. © Olga Povoroznyuk.*

in articulating and defending their interests in the process of planning a new large-scale infrastructure in Nome? To which extent do they still believe in the next big development plan?

Conclusions

So, how should we understand the specific constellation of reactions to this infrastructural megaproject? Given the contentious historical trajectory of Nome, it is not unreasonable to assume that we might need to look for distinct explanatory models for the settler and the Indigenous communities. Audra Simpson (2014, 2017) has introduced the notion of "refusal" to deal with specific ways of interacting with the (post)colonial state. While Simpson used refusal in the context of "consent" and "recognition," her terminology might be productive in our case as well. At least on part of the Indigenous elite, the supposed disengagement with the topic should not be read as disinterest. Could it be a refusal to engage again in a game that is stacked against them?

At the same time, we saw that many "white" residents of Nome share similar concerns with the Indigenous people. The prospect that the new infrastructural development will inevitably reduce subsistence opportunities, and the port construction process will put pressure on housing and social services, will affect both Indigenous residents and non-Indigenous settlers.

One of the themes appearing in many chapters of this volume is the question of agency vis-à-vis infrastructures. Our fieldwork in the summer of 2022 provided a snapshot of how particular individuals and groups related to the project in Nome. In analyzing it, we need to understand that we are dealing with a lengthy process of social and environmental relations determined by political and economic power. In a way, the port expansion project cannot be disentangled from Nome's history since the days of the gold rush (or even before) and what happened since. This processual perspective helps us in understanding the historical positionality of individual agency. At the same time, it offers the possibility to anticipate other constellations of social and environmental relations in the future.

As Thomas Owen (2008) has convincingly argued, Russian xenophobia and American distrust in dealings across the Bering Strait are at least as old as the Nome gold rush, that is they precede the Soviet period of Russian history. Not surprisingly, relations did not improve during most of the over seventy years, when the United States and the Soviet Union stood on opposite ends of an ideological showdown. Relations started to improve during the final years of socialist Russia, that is during the late 1980s, and things continued to look optimistic for most of the 1990s. Without claiming any clear causal correlation, one could argue that the political and quotidian relations across the Bering Strait have progressively deteriorated since the end of the 20th century, when Vladimir Putin came to power. While certain regional forms of coordination and collaboration continued until recently (e.g., in the form of the Eskimo Walrus Commission), high-level politics in the Bering Strait region have not been positive for a while.

The increased use of what English and Dutch seafarers of the 17th century used to call the Northeast Passage (see Introduction) —and what Russia refers to as the Northern Sea Route and China envisions as the Polar Silk Road—falls within the last 15 years, that is a period of distant but more or less collegial relations across the Bering Strait. Political and geographic conditions in the Bering Strait region did not make it necessary to have close collaboration for having this maritime passage to work. On the one hand, shipping lanes on the western side of the strait are under full political control of the Russian Federation, while the absence of deep-water ports on the US side—and the presence of the natural deep-draft harbor of Provideniya on the Russian side—excluded any necessity for using the eastern side of Bering Strait.

While the city of Nome has a long history of boom-and-bust cycles (Pearson, Rhoades, and Lewis 1990), the peak of the most noticeable boom—the gold-rush—occurred more than 120 years ago. The settler colonial society of Nome seems to have been in a state of constant anticipation of the next big thing to come. The Indigenous community/ies of Nome, on the other hand, seem to perceive the time as being ripe to counter the long history of non-Indigenous development plans in the region with ideas that focus on subsistence and Native livelihood.

In the end, we might be dealing with a particular "moment of stagnation" in a lengthy process that is characteristic of any large-scale infrastructure project, from the first ideas to implementation to the unavoidable decay. In the case of the Nome port expansion, a decision has been made, and—through the Russian invasion of Ukraine—this decision has gained enormous geopolitical significance ever since. In a way, everyone in Nome understands now that the port expansion plans are without alternatives. There is limited use of protesting plans that will be implemented one way or the other. Similarly, this is not the time for basking in dreams of potential economic and other benefits. At this point, the Nome port expansion project does seem to engender neither promises nor fears. It seems to be perceived more like a necessity that is a project that cannot be negotiated. It might well be that a more critical engagement with the project has to wait till the beginning of actual construction and, subsequently, the operation of the expanded port. While the new port has taken on the status of being unavoidable, the specifics of how it will be built and operated might offer new opportunities for engagement.

Thus, answering our research question, we can conclude that some of the local concerns, and especially Indigenous ones, regarding the rationale, social, and environmental costs, and the distribution of the potential benefits of the port expansion project have been overshadowed by national interests and plans. The lack of dialogue between the local and the national levels is even more visible today, in the context of the rapidly changing geopolitics of the Arctic. The historically rooted polarization of the local community between Indigenous and settler interests is another challenge for finding a way forward toward sustainable community development, while dealing with the advent of a large-scale infrastructure complex, largely planned and financed by external actors.

Epilogue

We conducted fieldwork in Nome in the summers of 2022 and 2023, but this article reflects primarily on the situation in 2022. Given our theoretical conviction that human-infrastructural relations are processual, we chose to focus on the specific constellations of actors and interests we experienced in the summer of 2022. In the summer of 2023, the dis-engagement described above had largely given way to more pronounced positions pro and con—primarily the latter—vis-à-vis the port expansion.

Finally, in October 2024, the Nome seaport expansion was suspended due to withdrawal of federal funds. The US Army Corps of Engineers, the main agency responsible for the port modification, explained the decision by high project costs. In February 2025, news about a renewed opening of bids for the port of Nome construction phase was publicly announced. The original construction project was redesigned with the Phase One split into parts A and B, in order to get the process going on at a smaller and more affordable scale (Lionas 2025). It will be interesting to see how things develop from here on and how the residents of Nome will deal with the stops and goes of the process.

Olga Povoroznyuk is a Postdoctoral Researcher and a Lecturer at the Department for Social and Cultural Anthropology, University of Vienna, and a member of the Austrian Polar Research Institute. She has been researching and widely publishing on the issues of infrastructure, development, identity, ethnicity and Indigeneity, postsocialism, and climate change in Siberia, Alaska, Northern Norway, and other Arctic and remote regions. Currently, she is leading the Vienna team on the Belmont Forum project ARCA about biocultural heritage and climate adaptation and has been coordinating and conducting research on transport infrastructure (particularly, seaports) and sustainable communities within the ERC project InfraNorth.

Peter Schweitzer is Professor at the Department of Social and Cultural Anthropology at the University of Vienna and Professor Emeritus at the University of Alaska Fairbanks. He is past president of the International Arctic Social Sciences Association (IASSA) and served as director of the Austrian Polar Research Institute (APRI) from 2016 to 2020. His theoretical interests range from the history of anthropology to human–environmental interactions, including the social entanglements with infrastructure and climate change; his regional focus is the circumpolar North. He has published widely on these issues and has been leading the ERC Advanced Grant InfraNorth.

Acknowledgements

This chapter is based on research funded, primarily, by the European Research Council (Project "InfraNorth," PROJECT-ID: 885646). The final stages of the work on the manuscript were supported by the Belmont Forum, the Austrian Research Promotion Agency FFG and the Austrian Ministry of Education and Research BMBW (Project "ARCA", project ID: 999909343/52036292) and the EU Horizon Europe HORIZON-CL6-2023-ZEROPOLLUTION-01 project "Illuq" (ID: 101133587). We would like to express our sincere gratitude to various local organizations and interlocutors, and in particular to Indigenous residents of Nome, for their time, engagement and openness to share their stories and perspectives. Last but not least, we are thankful to the editors of this volume for their enormous scrupulous work, including discussions of the chapters in a workshop format, their generous feedback and guidance throughout the editorial process.

Notes

The draft of this chapter was submitted to the volume in February 2023, went through several rounds of revisions and was last updated in April 2025.

1. KC (alias). Interview with the authors, Nome, July 2022.
2. KSZ (alias). Interview with the authors, Anchorage, July 2022.
3. OK (alias). Interview with the authors, Nome, July 2022.

References

Abbott, Richard J., and Christian Brochmann. 2003. "History and Evolution of the Arctic Flora: In the Footsteps of Eric Hultén." *Molecular Ecology* 12: 299–313.

Anand, Nikhil, Akhil Gupta, and Hannah Appel, eds. 2018. *The Promise of Infrastructure*. Durham, NC: Duke University Press.

Arctic Commitment Act. 2022. *A Bill to Enhance United States Standing as an Arctic Nation by Facilitating Greater Maritime Accessibility, Strong Trading Partners, and Reliable Infrastructure*. S. 4736, 117th Congress. Introduced in Senate on August 2, 2022. Sponsor Senator Lisa Murkowski (R-AK), Co-Sponsor Senator Angus King (I-ME). https://www.congress.gov/bill/117th-congress/senate-bill/4736. Last retrieved on August 7, 2025.

Bennett, Mia M., Scott R. Stephenson, Kang Yang, Michael T. Bravo, and Bert De Jonghe. 2020. "The Opening of the Transpolar Sea Route: Logistical, Geopolitical, Environmental, and Socioeconomic Impacts." *Marine Policy* 121, 104178.

Boas, Franz. 1905. "The Jesup North Pacific Expedition." *Proceedings of the International Congress of Americanists*, thirteenth session held in New York in 1902: 91–100.

Bruemmer, Fred. 1977. "Life on a Cold Rock." *Natural History* 86(3): 54–65.

Carse, Ashley, Jason Cons, and Townsend Middleton. 2018. "Preface to the Issue on 'Chokepoints.'" *Limn* 10: 2–3.

Chaussonnet, Valérie. 1995. *Crossroads Alaska: Native Cultures of Alaska and Siberia*. Washington, DC: Arctic Studies Center, National Museum of Natural History, Smithsonian Institution.

Cole, Terrence. 1984. *Nome: City of the Golden Beaches*. Anchorage: Alaska Geographic Society.

_____. 1990. "The Bridge to Tomorrow: Visions of the Bering Strait Bridge." *Alaska History* 5(2): 1–15.

_____. 2001 [1992]. "Jim Crow in Alaska: The Passage of the Alaska Equal Rights Act of 1945." In *American Nations: Encounters in Indian Country, 1850 to the Present*, ed. Frederick Hoxie, Peter Mancall, and James Merrell, 312–28. New York: Routledge.

Crouch, Maria Christina, and Jordan P. Lewis. 2023. "Alaskan Native Quality of Life: Culture, Rurality, and the Sacred." In *Rural Quality of Life*, ed. Pia Heike Johansen, Anne Tietjen, Evald Bundgård Iversen, Henrik Lauridsen Lolle, and Jens Kaae Fisker, 94–110. Manchester: Manchester University Press.

Demuth, Bathsheba. 2019. *Floating Coast: An Environmental History of the Bering Strait*. New York: W. W. Norton.

Eldridge, Kelly. 2014. "The Snake River Sandspit (NOM-146): A Late Western Thule Site in Nome, Alaska." *Alaska Journal of Anthropology* 12(1): 53–69.

Fitzhugh, William W., and Aron Crowell, eds. 1988. *Crossroads of Continents: Cultures of Siberia and Alaska*. Washington, DC: Smithsonian Institution Press.

Harvey, Penny, and Hannah Knox. 2012. "The Enchantments of Infrastructure." *Mobilities* 7(4): 521–36.

Haycox, Stephen. 2002. *Alaska: An American Colony*. Seattle: University of Washington Press.

Hays, Otis Jr. 1996. *The Alaska–Siberia Connection: The World War II Air Route*. College Station: Texas A & M University Press.

Hoffecker, John F., and Scott A. Elias 2007. *Human Ecology of Beringia*. New York: Columbia University Press.

Hopkins, David M. ed. 1967. *The Bering Land Bridge*. Stanford: Stanford University Press.

Hultén, Eric. 1937. *Outline of the History of Arctic and Boreal Biota during the Quaternary Period*. Stockholm: Thule.

Humpert, Malte. 2022. "International Shipping on Northern Sea Route Collapses as Foreign Companies Stay Away." *High North News*, September 12, 2022. https://www.highnorthnews.com/en/international-shipping-northern-sea-route-collapses-foreign-companies-stay-away. Last retrieved on August 7, 2025.

———. 2023. "Russia Says Northern Sea Route to Transport 270m Tons by 2035." *High North News*, April 17, 2023. https://www.highnorthnews.com/en/russia-says-northern-sea-route-transport-270m-tons-2035. Last retrieved on August 7, 2025.

Huntington, Henry P., et al. 2015. "Vessels, Risks, and Rules: Planning for Safe Shipping in Bering Strait." *Marine Policy* 51: 119–127.

Huntington, Henry P., Sarah Bobbe, Andrew Hartsig, Erika J. Knight, Alexey Knizhnikov, Alexander Moiseev, Olga Romanenko, Melanie A. Smith, and Benjamin K. Sullender. 2019. "The Role of Areas to be Avoided in the Governance of Shipping in the Greater Bering Strait Region." *Marine Policy* 110, 103564.

Huntington, Henry P., Julia Olsen, Eduard Zdor, Andrey Zagorskiy, Hyoung Chul Shin, Olga Romanenko, Bjørn Kaltenborn, Jackie Dawson, Jeremy Davies, and Erin Abou-Abbsi. 2023. "Effects of Arctic Commercial Shipping on Environments and Communities: Context, Governance, Priorities." *Transportation Research Part D* 118.

Johnson, Kirk R., Leo J. Hickey, and Cynthia Adams Hoover, eds. 1991. *Crossroads of Continents: The Material Culture of Siberia and Alaska*. Washington, DC and New Haven, CN: Smithsonian Institution and Yale University.

Jones, Preston. 2007. *Empire's Edge: American Society in Nome, Alaska, 1898–1934*. Fairbanks: University of Alaska Press.

Kendall, Laurel, and Igor Krupnik, eds. 2003. *Constructing Cultures Then and Now: Celebrating Franz Boas and the Jesup North Pacific Expedition*. Washington, DC: Arctic Studies Center, National Museum of Natural History, Smithsonian Institution.

Kettle, Nathan P., John E. Walsh, Lindsay Heaney, Richard L. Thoman Jr., Kyle Redilla, and Lynneva Carroll. 2020. "Integrating Archival Analysis, Observational Data, and Climate Projections to Assess Extreme Event Impacts in Alaska." *Climate Change* 163: 669–87.

Knight, Greg. 2023. "Nome Common Council Receives Update on Arctic Deep Draft Project." *knom.org*. April 11, 2023. https://knom.org/2023/04/11/nome-common-council-receives-update-on-arctic-deep-draft-project/. Last retrieved on August 7, 2025.

Lionas, Anna. 2025. "New Round of Bids to Open for Port of Nome Expansion." *The Nome Nuggett*, February 14, 2025. https://www.nomenugget.com/news/new-round-bids-open-port-nome-expansion-0. Last retrieved on August 7, 2025.

Loewi, Peter. 2022. "Arctic Commitment Act Surprised Region." *The Nome Nugget*, September 1, 2022. http://www.nomenugget.net/news/arctic-commitment-act-surprised-region. Last retrieved on August 7, 2025.

Milligan, Sean. 2022. "Nome Port Expansion Project Needs Additional Funding by November. " *KNOM Radio Mission*, May 4, 2022. https://knom.org/2022/05/04/nome-port-expansion-project-needs-additional-funding-by-november/. Last retrieved on August 7, 2025.

Minnich, Rick. 2022. *The Strait Guys*. Documentary film. Germany.

Monet, Jenni. 2023. "Nome Mayor Casts City as Arctic Partner at Arctic Encounter Gathering." *The Nome Nugget*, April 15, 2023. http://www.nomenugget.com/news/nome-mayor-casts-city-arctic-partner%C2%A0-arctic-encounter-gathering. Last retrieved on August 7, 2025.

Owen, Thomas C. 2008. "Chukchi Gold: American Enterprise and Russian Xenophobia in the Northeastern Siberia Company." *Pacific Historical Review* 77(1): 49–85.

Panahi, Roozbeh, Adolf K.Y. Ng, Mawuli Afenyo, Yui-yip Lau. 2021. "Reflecting on Forty Years Contextual Evolution of Arctic Port Research: The Past and Now." *Transportation Research Part A* 144: 189–203.

Pearson, Roger W., Edwin M. Rhoades, and Carol E. Lewis. 1990. "Alaska's Multibooms." *Technical Report, School of Agriculture and Land Resource Management, Agricultural and Forestry Experiment Station*. Fairbanks: University of Alaska Fairbanks.

Povoroznyuk, Olga. 2022. "Soviet Infrastructure in the Post-Soviet Era? Building a Railroad and Identity along the Baikal–Amur Mainline in East Siberia." Ph.D. dissertation, University of Vienna.

Ramseur, David. 2017. *Melting the Ice Curtain: The Extraordinary Story of Citizen Diplomacy on the Russia-Alaska Frontier*. Fairbanks: University of Alaska Press.

Rodgers, Dennis, and Bruce O'Neill. 2012. "Infrastructural Violence: Introduction to the Special Issue." *Ethnography* 13(4): 401–12.

Rogers, Karla. 1998. "Discrimination, a Reality in Alaska." In *Authentic Alaska: Voices of Its Native Writers*, ed. S. B. Andrews and J. Creed, 132–33. Lincoln: University of Nebraska Press.

Romanenko, Olga. 1994. *Mammals of Central Beringia: A Taxonomic List in English and Russian*. National Park Service, Beringian Heritage International Park Program; National Audobon Society; All Russian Institute of Nature Conservation and Reserves.

Rubano, Catherine. 2021. "Congress Authorizes Deepwater Port in Nome." *Alaska Public Media*, January 14, 2021. https://alaskapublic.org/2021/01/14/congress-authorizes-nome-deepwater-port-project/. Last retrieved on August 7, 2025.

Russell, Emily. 2015. "Visa-Free Travel to Russia Reinstated for Eligible Alaska Natives." *Alaska Public Media*, August 10, 2015. https://alaskapublic.org/2015/08/10/visa-free-travel-to-russia-reinstated-for-eligible-alaska-natives/. Last retrieved on August 7, 2025.

Schneider, William S. 2012. *On Time Delivery: The Dog Team Mail Carriers*. Fairbanks: University of Alaska Press.

Schweitzer, Peter P. 1997. "Traveling Between Continents: Native Contacts Across the Bering Strait, 1898–1948." *Arctic Research of the United States* 11: 68–72.

Schweitzer, Peter P., and Evgeniy Golovko. 1995. "Traveling Between Continents: The Social Organization of Interethnic Contacts Across Bering Strait." *The Anthropology of East Europe Review* 13(2): 50–55.

Schwing, Emily. 2023. "As Arctic Traffic Increases, Nome Grapples with Its Future. 'It's Like a Highway Going Right Past Us.' " *KYUK*, March 1, 2023. https://www.kyuk.org/science-and-environment/2023-03-01/as-arctic-shipping-traffic-increases-nome-grapples-with-its-future-its-like-a-highway-going-right-past-us. Last retrieved on August 7, 2025.

Simpson, Audra. 2014. *Mohawk Interruptus: Political Life Across the Borders of Settler States*. Durham, NC: Duke University Press.

———. 2017. "The Ruse of Consent and the Anatomy of 'Refusal': Cases from Indigenous North America and Australia." *Postcolonial Studies* 20(1): 18–33.

VanderBerg, Justin D. 2018. "Optimal Arctic Port Locations: A Quantitative Composite Multiplier Analysis of Potential Sites." *Polar Geography* 41(1): 55–74.

West, Frederick Hadleigh, ed. 1996. *American Beginnings: The Prehistory and Palaeoecology of Beringia*. Chicago: The University of Chicago Press.

Conclusion

The Resonance of the Unbuilt

Matthew Magnani and Natalia Magnani

What is the impact of the unbuilt on human agency? When realized, infrastructural projects move goods, people, and practices, reshaping cartographies of exchange and the social interactions that accompany them. For the nation-states and sovereign bodies that conceive them, infrastructural projects reflect and shape aspirations for prosperity and power. For those who build the structures or reside in their path, they translate into shades of displacement, opportunity, or ambivalent presence.

An anthropology of the built environment reveals how infrastructural projects leave long-term marks on the landscape, shaping people's actions and their movement well beyond periods of construction. Even Stone Age settlement patterns transformed the fecundity of contemporary landscapes. Old walls, bridges, and roads inform human movement and have a gravity that attracts greater growth. Game trails that long invited human traffic were turned into dirt roads, to be paved for cars and paralleled by electric lines and railroad tracks. Decisions made thousands of years ago influence the way people aggregate, socialize, and visualize their contemporary social worlds along long-forgotten paths.

The built landscape, often visible at the surface and durable intergenerationally, has occupied anthropological interest for generations. Monumental architecture is as much a magnet for invading armies as it is for archaeologists. Yet for every finished project, many more are left on the drafting table, partially built, or lingering in the imagination. Fierce resistance to building can stop plans in their tracks. Shifting political winds, a changing climate, or dried up sources of funding aggravate one another, rendering building projects defunct. To understand how infrastructure functions socially, it is productive to look beyond what is finished, to what is imagined and anticipated.

Silk Roads of Practice and Anticipation

Building projects may cut across vast distances, directed by sovereigns that mobilize labor. In other cases, infrastructure is not the result of a unified vision but paved by a constellation of smaller-scale actors responding to broader exchanges and movements of people. Such was the case with the Silk Road in antiquity. A figurative conduit of trade that spanned land and water, was named centuries after the fact by connecting artifacts, texts, and later genetic markers from plants, animals, and people. Day-to-day participants in these networks would have had more narrowly defined experiences, rarely connecting their products to the final markets where they ended up being interred. Historically conceived and projected into the past, these old trade routes nevertheless left a weight on the landscape that is immense, hinting at diverse patterns of ancient trade.

The imagination of past infrastructure conjures future networks of exchange. Any framing of the built world of the past must therefore be contextualized alongside contemporary desires. The Silk Road was unified by 19th-century scholars, reflecting interest in greater exchange between east and west. Just as broad trade networks were subsumed under the name "Silk Road," today the terminology is being revived to summon a loosely connected but extensive network linked to melting sea routes in the Arctic. Even if the Silk Road never existed as a cohesive unit historically, ancient networks are unified to promote development and economic growth. Launching from hundreds of years of searching for shortened sea routes, nations vie for control over melting passages and commerce, as they enact their own visions of sovereignty and prosperity. An analysis that follows the anticipation of these routes reveals how local agencies emerge alongside these imagined pasts, evoked in contemporary politics to connect, challenge and secure future transit.

Ethnographies of the Unbuilt

Evoking Silk Roads as a heuristic to examine the expansive networks that are anticipated along a loose pathway of imagined building projects in the Arctic, the case studies in this volume attest to the spaces created by infrastructural projects as domains of potential, risk, and anxiety for the people and communities along their projected paths. In Sápmi, port infrastructure and associated business zoning emerge as sites of both opportunity and apprehension for a flourishing Sámi business, which create new attachments to place and identity (Magnani and Magnani, this volume). In China, the Ewenki contribute to establishing a nation in near-Arctic terms, creating space for Indigenous agency to come into dialogue with state building projects (Fraser, this volume). In northeast Siberia, ghosts of Soviet development policies intersect with plans for Chinese and other foreign projects that often fail to materialize—from bridges to refineries, fiberoptic internet connections, and deep-sea ports (Stammler, Ivanova, and Vitebsky, this volume). Each of the chapters reveal a patchwork of agencies that emerge in the footprints of not-yet-realized infrastructure. Anticipation resonates in material worlds without the scaffolding of the physical structures of ports, piers, rails, airstrips or roads.

Connecting a range of ethnographic perspectives from across these sites promises to increase a depth of understanding on the role of anticipation associated with large-scale building projects. Rooted in multi-temporal perspectives and drawing on disciplines from anthropology, history, archaeology and human geography, ethnographies of the unbuilt articulate and reveal a patchwork of agencies associated with anticipation. Such approaches illuminate the degree to which global infrastructural development resonates with the desires of local actors, suggesting that where large-scale projects are compatible with local visions, the largest footprint of the unbuilt may result. Where state-ordered building projects are poorly aligned with local desires, the unbuilt is less likely to leave footprints or transform practice. This highlights the relative importance of local agency in the reshaping of landscapes through times of uncertainty and anticipation.

Turning ethnographic vantages on the unbuilt is particularly salient in the Arctic, where state interests manifest beyond traditional forms of territorial sovereignty. Trade and security goals play out under ice packs as flags are planted miles under the water on the ocean floor. Icebreakers and cargo ships push the boundaries of what is navigable, an extension of centuries of attempts to shorten trade routes through frozen waters. Assertions of sovereignty are made through plans for connectivity that may never be realized, yet tangibly and affectively extend the presence of state actors into new zones. In these places, grandiose visions of national security, transport infrastructure, and resource extraction may not necessarily come to fruition materially but resonate in other ways.

Routes Forward

Through the 21st century, infrastructural imaginaries across the Arctic will continue to be inextricably linked to climate change, and the patterns of melting ice and rising sea levels that open and shut new maritime corridors and ports of exchange. Cargo ships will explore new paths at new times of year, less dependent on icebreakers. These transformations will open new physical trade routes and decrease fuel consumption but also drive the increased consumption of raw materials and the production of goods. Ports will open, close, and relocate in response, transforming not only the trajectories of shipping corridors but the routes that materials take over land. As the chapters in this volume demonstrate, local actors may emerge to cultivate the shadows of these infrastructures, not yet realized, both coping with and capitalizing on anticipation.

Intersecting with global transformations and the impacts of climate change, equally seismic political upheaval will contribute to a reorganization and hardening of state borders, catalyzing and constraining where Arctic and near-Arctic states and their citizens focus their desires. Just through the process of editing this volume, changes in geopolitics and global conflict have co-opted plans for commerce, yielding to an era of military preparedness and economic sanctions over free trade. Plans for routes that threatened to cut across reindeer pastures as part of the Belt and Road Initiative have been overtaken by NATO. Discourse on Indigenous displacement is increasingly clouded by assertions of national security. Considering the inverse, looking back to the historical record reminds us that major road and rail networks built to ferry troops may later facilitate travel, tourism, and commerce. With mercurial political shifts, how major building projects will change shape, not only on the ground but in the imaginations of Arctic residents, will continue to prove a valuable site to understand the influence of anticipation on practice.

How will the specters of Arctic Silk Roads continue to resonate along their lengths, realized in some places and abandoned as ideas in others? Infrastructures will continue to alter landscapes through the pouring of new concrete and clearing of trees, but in many more cases they will affectively reorganize people around pathways that never materialize. An anthropology of the unbuilt calls scholars to be aware of the ghosts of unrealized projects, where no physical structures shape movement. Road systems or ports that may have only been constructed in discourse may still organize and influence practice long after their plans are forgotten. In years to come they may continue to inform action where they were first envisioned, without any foundations except for the anticipation of earlier actors. Looking beyond the built, we expect the echoes of the unrealized to resonate, however visible, in the landscapes and ethnographies of the future.

Matthew Magnani is Assistant Professor of Anthropology and Climate Change at the University of Maine, and a researcher on the "Arctic Silk Road" project based at UiT The Arctic University of Norway. He has published dozens of articles on material culture past and present, in journals such as *American Antiquity* and *Current Anthropology*, and he is the co-author of *The Craft of Belonging: Material Culture and Social Boundaries in Sápmi* with Natalia Magnani (University of Toronto Press).

Natalia Magnani is Assistant Professor at the Department of Anthropology and Climate Change Institute at the University of Maine. She leads the Norwegian Research Council project "Arctic Silk Road" based at UiT The Arctic University of Norway, whose participants have collaborated in shaping this volume. She has published numerous articles on the Sámi regions, including in *Current Anthropology* and *American Anthropologist*, and she has two books forthcoming—*Silent Knowledge: Reviving Sámi Practice* (University of British Columbia Press) and *The Craft of Belonging: Material Culture and Social Boundaries in Sápmi* with Matthew Magnani (University of Toronto Press).

References

Fraser, Richard. (this volume). "Imagining the Arctic in Northeast China: Ethnic Minorities and the State along the Polar Silk Road." Oxford and New York: Berghahn Books.

Magnani, Natalia, and Matthew Magnani. (this volume). "Economies of Ghost Infrastructures in Arctic Norway." Oxford and New York: Berghahn Books.

Stammler, Florian, Aytalina Ivanova, and Piers Vitebsky. (this volume). "Vast spaces and Inconsistent Anticipations in the Northeast Siberian Arctic." Oxford and New York: Berghahn Books.

Index

www.ingramcontent.com/pod-product-compliance
Ingram Content Group UK Ltd.
Pitfield, Milton Keynes, MK11 3LW, UK
UKHW021836150926
8656IPUK00004B/21

9 781836 953081